ASTROBIOLOGY

AND THE SEARCH FOR LIFE IN OUR UNIVERSE

A STUDY OF SPACE, CHEMISTRY, HISTORY, EVOLUTION, LIFE, EXOPLANETS, EXTRATERRESTRIALS, AND CIVILISATIONS

KESHAV SANCREDI

Copyright © 2024 by Keshav Sancredi

Published in Malaysia by:
Keshav Sancredi

First paperback edition, published in 2024

All rights reserved.
No part of this book may be used or reproduced without the written consent of the author.

ISBN: 9798325390876

To Mama and Ba,
For everything.

To Master Bathi,
For teaching me.

To Uncle Kana and Aunty Joanne,
For the fond memories.

To Lehn, Tylann, and Baby,
For being the best brothers ever.

Contents

Prologue — 1
    The Sky is Falling — 2
    Of Literature and Movie Magic — 4
    The Cosmic Calendar — 5
    The Past, Present, and Future of Science — 7
    A Personal Insight — 11
    The Coalescence of Science — 14

Chapter 1: Stars, Planets, and Everything in Between — 21
    Genesis — 22
    Let There Be Light — 25
    A Hitchhiker's Guide — 29
    The Great Abyss — 35
    Of Everything Spherical, or Otherwise — 40
    The Circle of Life — 46
    The Lightbulbs of Our Universe — 58
    Your Home Is Here — 62
    The Great Merry-Go-Round — 66
    Gallery — 72

Chapter 2: Chemistry of the Universe — 81
    Elementary, My Dear Watson — 82
    May the Forces Be With You — 88
    Angels and Demons — 99
    Genesis of the Elements — 104
    In the Heart of a Star — 107
    Life From Death — 112
    Making Superheroes and Supervillains — 115
    The Highs and Lows of Energy — 121
    Gallery — 129

Chapter 3: History of Earth — **145**
    Aetas Terrae — 146
    Hell on Earth — 151
    The Calm After the Storm — 156
    Laying the Groundwork — 158
    Let There Be Life — 162
    Welcome to Jurassic Park — 169
    Class Mammalia — 177
    Radioactive Rock — 183
    One in a Septillion — 191
    Gallery — 203

Glossary — **229**

Further Reading — **251**

Index — **255**

Prologue

"We are all connected. To each other, biologically. To the Earth, chemically. To the rest of the universe, in time."

Neil deGrasse Tyson

The Sky is Falling

In the early 1900s, the Earth was invaded by aliens, and humanity came within a hair's width of extinction.

For obvious reasons, there is not a person alive today that can accurately recall what exactly took place. But through the hard work of historians and archeologists, we can speculate that these intruders were from our very own backyard, the average-sized red planet next door. These 'Martians' came in their thousands, with goliath machines armed to the teeth with death rays, laser beams, and chemical warfare.

Miraculously, in what can be coined the ultimate deus ex machina of the century, the Martians were seemingly killed by an onslaught of Earthly pathogens, from which they had no immunity. Silly Martians. You would think that before one masters the ability to control lasers and heat rays, one would first protect oneself from the common cold. However, they left an undeniable mark on humanity that is still evident today. Infact, aliens have been leaving their John Hancock on Earth since time immemorial.

Erich von Däniken, a Swiss author, has penned numerous books which make claims of extraterrestrials on early human culture, including the bestselling, *Chariots of the Gods?*. He postulates that some ancient structures and artefacts appear to reflect more sophisticated technological knowledge than is known or presumed to have existed at the times they were manufactured.

Such artefacts include the Pyramids of Giza, the Nazca Lines, Stonehenge, and the Moai of Easter Island. The book also suggests that ancient artwork throughout the world can be interpreted as depicting astronauts, air and space vehicles, extraterrestrials, and complex technology, far before their time.

Do you really think that the Egyptians built the Pyramids themselves without any extraterrestrial support? That Stonehenge just happens to be a random assortment of huge and weighty rocks in the middle of nowhere? How do you think the Mayans and Vedas were able to obtain their vast knowledge of Mathematics, constellations, and medicine? And what of the

Prologue

Nazca Lines in Peru? A potential method of extraterrestrial communication?

Now, obviously all of this is mere speculation, and has no basis in fact and evidence, which are the main principles that science is founded on. However, the concept of questioning the existence of alien life, and if it has affected us and might do so in the future, should not be put down.

The presence and influence of potential alien life in our past has been a sort of 'hot and cold' topic in society. On one hand, the average History channel viewer tunes in to Ancient Aliens and is whisked away to a world of pseudoscience and conspiracy theories. On the other hand, anyone with a basic background of science and history can identify these fantastical claims for what they are. A mere attempt to grasp the attention of an audience that can be easily swayed by sketches of aliens and a few conveniently-timed coincidences. We will return to these conspiracy theories later on, when we discuss UFOs and if there really are extraterrestrials living among us. And no, of course Martians didn't invade Earth (don't rule it out just yet though! As we shall see later on). I owe that bit of imagination to H.G. Wells, who set the tone for modern sci-fi epics and novels with his book *The War of the Worlds*.

So what was the point of this misleading information? Merely to demonstrate that humans have, for millennia, questioned the possibility that we are not the only life forms in the universe. This curiosity spans all the way back to ancient Greece, with the works of Democritus and Epicurus. We will get into this topic in far more detail when we discuss the possibility of life on exoplanets, but for now it fits the purpose of showing that we, as humans, have been on a lifelong quest as a species to discover the true meaning of our existence and if there really is 'alien life' in the universe. Surely we can't be all alone in the void of space?

Speaking of this scientific curiosity, the ability of an everyday society to absorb and process new information on scientific matters, is actually a testament to the age we live in. We have evolved from science being the privilege of a select few, to an entire world having access to almost any information in a matter of seconds. The world has science at their fingertips. This has made it far easier for authors, teachers, and people such as myself, who aim to spread the wonders of science, to output our knowledge into the world, for everyone to access. And the main access point for the layman, is in literature and on the silver screen.

Of Literature and Movie Magic

On average, in the United States, around 500 movies are released each year. Of these 500, the science fiction genre consists of around 10%. This might not seem too significant, but there has been a steady increase in not only the number of sci-fi movies produced each year, but also the percentage of revenue these movies make, accounting for around 30% - 40%.

The beauty about sci-fi movies is the sheer range applicable to it. This is once again a testament to the countless aspects of science that can be explored. Want to know how biologically tweaked humans and animals could function in society? *Jurassic Park* and *The Fly*. Wondering about the future of human civilization? *Blade Runner* and *Minority Report*. Interested in space exploration and planet colonisation? *Interstellar, Dune* and *The Martian*. So you see, there is a groundbreaking sci-fi movie for almost every conceivable topic in science, and this book will explore almost all of these topics and how they all fit in the jigsaw of life.

Away from the silver screen, science fiction has been prominent in literature for more than 2 centuries. The stories of Jules Verne, Mary Shelley, and H.G. Wells, to name a few. Each of these writers used different elements of science fiction in their novels, to portray incredible worlds and technologies that astounded the people of the time.

This tradition of constantly upping and broadening the imagination of writers has produced some of the most acclaimed authors of the past century. From the robots and galactic empires of Isaac Asimov, to the dystopian worlds of Philip K. Dick, and across the various planets and moons from the works of Arthur C. Clarke, even to modern day adventures

Prologue

by Andy Weir, the world of sci-fi in literature is evolving at breakneck speed, as new inventions and discoveries are made every day.

I will use examples from these sources to provide a mutual understanding that everyone is familiar with. Which one of us has not been in awe at the dazzling panoramas of space on the big screen? Have we not been sucked into the dystopian world of Panem through mere words on a page? These are topics and examples that we are all familiar with, and I will use them to show connections between the scientific topics to be discussed, and how it can be visualised through text or screen.

So whether it's the next Hollywood blockbuster set in a galaxy far, far away, or the soothing aura of a David Attenborough nature documentary, or even the next page turner from Michio Kaku or Stephen King, these methods of picking up new information on anything and everything science, which helps to broaden the curiosity and wonder of everyday people, provides a terrific method of visualisation to society.

This is visualisation in terms of scientific history, advancement, and possibility, but how will we visualise the sheer time scales in this book? For that, we turn to the Cosmic Calendar.

The Cosmic Calendar

What if I told you, that you have been on a 13.8 billion year journey, to reach this exact moment in time and space, where you happen to be reading this sentence? This precise moment that you exist in right now, has had 13.8 billion years of preparation. No, this is not another 'Martians and aliens' wind up, this is pure and simple fact.

But what is 13.8 billion years? To put that number into perspective, human beings or *homo sapiens* have only been 'alive' on Earth for 350,000 years. Back of the hand calculations tell us that humans have existed for only 0.00002% of time, since the birth of the universe.

Famed American astronomer and scientist Carl Sagan has illustrated this perspective of time in an even more startling manner, with his Cosmic

Calendar. First popularised in his book *The Dragons of Eden* and on his television show *Cosmos*, the Cosmic Calendar puts our place and time frame in the universe into a rather miniscule perspective. Essentially, if the Big Bang (the theoretical start of the universe) occurred on 00:01 am on the 1st of January, anatomically modern humans would only have come into existence on the 31st of December, at 11:52 pm.

Think of all the historical events that you have learnt of and studied in school. William the Conqueror, the Renaissance Era, the American and French Revolutions, the Battle of Waterloo, two World Wars, moon landings, assassinations, the inventions of the satellite, computer, and mobile phone. All these occurrences would have occurred on the final second, of the final minute, of the final hour, of the final day, in our Cosmic Calendar.

Throughout the course of this book, I will return to the Cosmic Calendar to provide us with a viewpoint in time, as we travel through the aeons, venture into foreign lands, and challenge our own place in the universe.

So how have we been on a journey for this long to reach this exact moment, which as we now know, is almost insignificant across the entire age of the universe? What is our connection with the stars, planets, and primordial matter of the universe? How can we even begin to perceive our place in this vast cosmos of innumerable possibilities, and what will become of us in the next 13.8 billion years? These are just some of the multitude of questions that we will ask and answer in this book.

Staying on the topic of Carl Sagan, (which we will return to frequently as he is a pioneer of astrobiology) he once said *"you have to know the past, to understand the present"*. This is a very powerful quote in terms of learning the history of scientific inquiry, and why we continue to ask questions and probe into the unknown. Why is all of this research important? What have we achieved so far? And how can we hope to impact humanity in the future? To answer these questions, we have to embark on a brief voyage in time to the past. 2300 years to be precise.

Prologue

The Past, Present, and Future of Science

Aristotle was an ancient Greek philosopher and polymath, who set the groundwork for the development of modern science. Born in Northern Greece, in the city of Stagira, Aristotle had a far from normal childhood. His father, Nicomachus, was personal physician to the King of Macedon, and gave young Aristotle a personal education in biology and medical information. Aristotle would become orphaned at 13, but his achievements and effect on the modern world have stood the test of time. He received a formal education at Plato's academy, and would even go on to tutor the son of King Philip, Alexander the Great. An entire encyclopaedia could be written on the works and wonders of Aristotle, but we will focus on his efforts in forming the scientific method.

Aristotle used inductions from basic observations to infer general principles, and from said principles, produced deductions which could be tried and tested. This same basic principle is still used today, almost 2,500 years later, at every laboratory in the world, and is our main method in proving the falsifiability of a theory or hypothesis.

Much can be said about the scholars who would go on to improve and reiterate the scientific method. From the scholars of the Muslim World, Ibn al-Haytham, Ibn Sina, and Ibn Rushd, to the renaissance men of Europe, Leonardo da Vinci, Copernicus, and Galileo, to the scientists of modern history, Newton, Einstein, and Curie. All of them learnt and formed their hypotheses from the scientific method developed by Aristotle. The history of science is another book in itself, but for the purpose of this book, we will turn to the scholars of astrobiology.

On the 1st of May 1875, in the village of Smolevichy, near Minsk, Gavriil Tikhov was born. Born into a railway employee's family, he constantly moved from place to place during his childhood. He completed his secondary education in Simferopol, the second largest city in the Crimean Peninsula. He writes in his memoir, *60 Years Near the Telescope*, that in the spring of 1892, after reading a couple books on astronomy and looking through a telescope for the first time, he irrevocably decided to be an astronomer. He has had two separate craters, one on the Moon and one on Mars, and an asteroid named after him. So how does Gavriil Tikhov fit into our history of astrobiology?

It was Gavriil who first proposed the term 'astrobiology' and was considered the father of astrobotany. One of the first astronomers who believed in the possibility of growing and sustaining life in space. Another associated term is xenobiology, coined by American science fiction writer Robert Heinlein in his book, *The Star Beast*. However, xenobiology is now used in a more specialised sense, referring to 'biology based on foreign chemistry', whereas astrobiology applies to all forms of life, whether foreign or not.

Scientists only started seriously considering the possibility of life on other planets during the height of the Space Age, in the 1950s and 1960s. The Russians launched their satellites, the Americans launched their probes, and the scientists on Earth waited with bated breath for any signs of extraterrestrial life. We will cover the history and future of man-made objects being sent into space in far more detail further on, when we discuss the SETI program.

The concept of life elsewhere in the universe was brought to the masses by the works of Carl Sagan. Among his countless contributions to science, he assembled the first physical messages sent into space, the Pioneer plaque, and the Voyager golden record, which were universal messages that could potentially be understood by any extraterrestrial intelligence that might find them. He contributed to the discovery of the high surface temperature of Venus (which we will touch on later, when we discuss exoplanets), hypothesised that the moons, Titan and Europa, might possess liquid components and oceans of water on their surface, and is one of the most cited SETI and planetary scientists ever.

Carl Sagan brought these concepts of extraterrestrial life to everyday people with his show, *Cosmos: A Personal Voyage*, as well as its companion book. He also penned the book and wrote the screenplay for the 1997 movie *Contact*, starring Jodie Foster and Matthew McConaughey,

Prologue

about a SETI scientist who finds evidence of extraterrestrial life and is chosen to make first contact.

In the present day, astrobiology is continuing to expand, but perhaps not at the rate envisioned by Carl Sagan or Gavriil Tikhov. Of course, any mission or research done in space will take years to be successful, not to mention the need for more and more sophisticated technology, but it is also stifled by growing costs and diminishing interest from public support. Understandably, society is more concerned with our planet here and now, and does not envision the vistas of Mars or Titan as possible worlds. This is a science that takes decades, if not centuries, to come to fruition, and any frustration on behalf of the scientist or public is not without cause. In such a bleak setting, why do we still persist in spending time, money, and effort, in searching for exoplanets, building telescopes, and exploring space?

In many ways, the job of an astronomer is a visualisation of one's childhood. I'm not saying that astronomers are children, what I mean is that every child on Earth has looked up at the heavens, seen the multitude of stars and lights in the sky, and stared in wonder at these beautiful creations. The concepts of nebulae, black holes, exoplanets, and nuclear fusion will surpass most children, who will forget these childhood curiosities once they are thrown into the ever burdening world. However, astronomers carry this curiosity into their education and careers, seeking to find answers and attempting to solve the big questions that the universe asks. Now, most people are content with our place in the universe, a point of view only limited to Earth. But for those of us who carry this childlike curiosity of wanting to know the secrets of the universe, space exploration and research provides a valuable and productive outlet.

Moreover, the effects to society from space travel and research have been, and will be, profound and long lasting. The first satellites were designed to study the space environment, and test the initial capabilities of man-made objects in Earth's orbit. Ultimately, this led to the development of satellites used for telecommunications, GPS, and weather forecasting, among others.

Exploring the Essence of Everything

The challenges posed by space travel have proved to be the driving force of innovation and efficiency. The cost of space launches forced engineers to redesign computers to be smaller, lighter, and more efficient, with the highest percent of dependability. No one wants a computer that shuts off on its own accord, 400,000 km from Earth. This eventually led to the creation of mobile phone cameras, by a team at the Jet Propulsion Laboratory (JPL), in 1990.

Every space mission needs outstanding and faultless communication, considering the fact that should anything go wrong, the astronauts are well and truly on their own. To make the lives of astronauts easier, and allow them to simultaneously cover more ground and carry out more tasks, NASA came up with the concept of 'wireless technology'. Yes, the airpods, wireless headphones, and telephones that we use everyday, stems from this basic need of astronauts in space.

From a medical standpoint, space has been a crucible of invention and innovation. NASA scientists, in an effort to protect astronauts from losing bone and muscle mass in the microgravity environment of space, helped a pharmaceutical company to test Prolia, a drug that today helps protect elderly people from osteoporosis. Aside from that, for anyone that has worn, or is wearing braces, one of the main components is actually a flexible but resilient alloy named nitinol, which was developed to enable satellites to spring open after being folded to fit inside a rocket.

These are just a handful of examples on how space travel and exploration has impacted, and can impact, our society. Throughout the course of this book, I will share a few more benefits, including the use of the International Space Station (ISS) for the future of crops, medicine, and robots, how the martian probes have improved the use of prosthetics and artificial limbs, how we can tap into the supply of asteroids for rare Earth materials and elements, as well as what we can expect from space travel in the future.

Prologue

A Personal Insight

The following section is a sort of autobiography on how my love for science blossomed, and why I happened to write this book. My inspirations, views, and what I hope to achieve and share with this work. However, if you would rather not read this section, feel free to skip to the next one on page 16.

When I was a child, two instances helped to shape my interest in science. Firstly, at the age of 5, my dad and I built a miniature model of the solar system (at the time consisting of 9 planets!). This was supposedly a project for one of my kindergarten classes, but I don't seem to recall any of the other students doing anything, so perhaps my dad just wanted an excuse to make something with me. Anyways, the model ended up being displayed on the wall of the classroom, where it still is today! (a testament to our engineering abilities, and superglue).

The second instance was when my parents got me a huge encyclopaedia on space. I reckon this is probably a more relatable occurrence to some of you who probably received the same book. The bigger words and terms made little sense to me at the time, but I was dazzled by the images and descriptions of celestial bodies. A spark had been ignited, and the rest, as they say, is history.

Fast forward 17 years later, and my love for science has grown and is applicable to everything that I'm passionate about. Take cooking for example. The science of cooking is known as molecular gastronomy, but it can be understood in far more simpler terms. The use of acids to tenderise proteins, heat to break down enzymes, the Maillard reaction which allows for that distinctive sweet caramelised taste. Temperature control is vital to kill any microorganisms but must be kept in check to prevent overcooking. How different types of cooking methods can change the taste, appearance, and texture of certain foods. All of these methods have scientific roots, and is just one example of how I use science in my day to day life.

Outside of the kitchen, I apply science (and particularly the works of Isaac Newton) in my Taekwon-do dojo. The ability to produce the maximum amount of power using the least amount of effort, using concepts of sine-wave and equilibrium to achieve a powerful kick or punch, and using an opponent's momentum against them in self-defence. Newton's third law applied to the backward motion used to produce the maximum amount of forward motion and force. Taekwon-do has a solid base and history in all these scientific methods, and is not simply a way to kick or block.

Away from the dojo, we move back to the world of pop culture references. After finishing high school, I was still undecided on what I wanted to achieve and do with my life. Thoughts of neuroscience, pathology, and marine biology floated around the ether. However, much like Gavriil Tikhov and how the stars and telescopes instantly inspired him, I too had a similar moment of eureka. And for this, I owe it to Chirstopher Nolan, and his movie *Interstellar*. *Interstellar* is universally acclaimed to be not just one of the best sci-fi movies of all time, but one of the best movies ever made in general. It has changed my life in ways that I can't fully express. It truly made me realise that space, and all of its limitless possibilities, has so much to offer us, in terms of exploration, discovery, technology, and changing the very fabric of spacetime. For those of you who are familiar with the movie, and want to know the science behind the wormhole, black hole, tesseract, and the planets, I highly recommend reading Kip Thorne's book *The Science of Interstellar*. I will use examples from this movie, and the methods of science that it portrays, further on in our story, as there are so many branches that are applicable to us in this study of life in the universe.

Another motion picture that changed my view on life and science is Ridley Scott's, *The Martian*. Based on the bestselling book by Andy Weir, it portrays the story of an astronaut stranded on Mars, and how he overcomes a plethora of challenges to make it back to Earth. This story is the foremost example of how life can exist and thrive on distant planets, in unimaginable conditions, as long as we have the innovation and imagination to overcome the obstacles faced. I will also come back to this story throughout the course of the book.

Prologue

They say that a movie is only as good as its screenplay, and how true that is. Behind every fascinating idea and script, is a dedicated, talented writer. Among the writers that have shaped my love for science, there are two in particular that I have to give a special mention to. The physicist Michio Kaku, and the author Isaac Asimov.

Michio Kaku is probably the most well known science author of our time, (I mean no offence to any other authors) with his works spanning a multitude of concepts and ideas. From colonising Mars, to the inventions of the next 100 years, to the impossibilities in physics, to his quest in finding the answer to the God Equation, his books have enlightened me to appreciate the beauty in science, and to show that even the most complex of situations and theories can be broken down in the simplest of terms, for everyone to understand. This has inspired me to follow suit and write a book of my own, that I hope in turn will inspire others.

Isaac Asimov, an American writer and professor of biochemistry, is truly a pioneer of the sci-fi trope, and has had an indirect hand in every sci-fi movie or book produced in the past 60 years. Born in Soviet Russia in the 1920s, to a family of Russian Jewish millers, Asimov has written or edited more than 500 books and scientific papers. Almost all concepts of science that we see in pop culture pays some homage to the works of Isaac Asimov. Whether it's robots, future space civilisations, space travel, or futuristic technology, Asimov has written about almost every sci-fi topic under the sun. His works have shown me the possibility of human imagination in making stories and characters out of a pure love of storytelling and science, and I hope to follow in his footsteps one day, with a sci-fi story of my own.

I encourage anyone interested in life, science, and space, to watch these movies and read books by these authors. It may change your view on matters as it has done so for me.

All of these examples have had a hand in me wanting to write this book. At its core, it's a story of how the bare constituent atoms of life have evolved, from a primordial soup of particles, to fully grown, conscious beings, on a habitable planet, that are now able to understand and explore their past and future. We will study almost every facet of science and how they all come together to explain our position in the universe, and what our future may look like. With that, let us take a closer look into what exactly this book has to offer.

The Coalescence of Science

Before we proceed any further, I should explain exactly how this book has been written, and why the prologue is unusually long. There are 12 main chapters that will be explored throughout the entire duration of this story. Each chapter will come in at around 70-100 pages, making the total length of the book some 1200 pages. Of course, it is absurd for a story of this length focusing on this subject matter to be published in one entire book. As a result, the entire book has been split into four volumes, of which this is the first. Keep in mind that each volume follows exactly where the previous one left off, meaning that this prologue covers the entirety of the 12 chapters. Vol.2 will immediately start with chapters 4 until 6, Vol.3 will start with chapters 7 until 9, and Vol.4 will start with chapters 10 until the epilogue which will once again cover all 12 chapters. In truth, this story is written as one continuous book, just split into four separate volumes for easier accessibility and reading convenience. So when you come across the phrase 'this book' it refers to the entire story encompassing all 12 chapters and 4 volumes, not just Vol.1.

The main ethos of this book is an attempt to reconcile the concepts, theories, and understanding of life (as we know it), its past, present, and future, with its place in the universe, amongst the countless stars and potential worlds across the cosmos. In fact, we can split the main book title into four sections, each of which will cover one volume. The 'Astro' part is in the first volume, with chapters on astrophysics and astrochemistry. The 'biology' part is in the second volume, with chapters exclusively on life; its origins, its evolution, and its chemistry. The 'Search for Life' is in the third volume, with chapters on exoplanets, signs of life, and SETI. 'Our

Prologue

Universe' is in the fourth volume, with chapters on civilisations, the future of life in millions and billions of years, and its eventual end.

All the facts presented will be supported by evidence and backed up by countless reports and theories done by scientists far more capable than myself. However, that will not deter us from having stimulating, eye-opening debates about the future of humanity, philosophical aspects of life, and the very nature of our universe. As it is said, the most groundbreaking ideas and theories stem from daydreaming and fantasising about the unknown.

There are four main sections in this book, each divided into three chapters. The chapters themselves are divided into a dozen or so subchapters to ensure a smooth and informative reading experience. The facts and theories will be presented in an orderly manner, with a definitive sequence of events, while making it accessible to understand.

The first section deals with the essence of everything in the universe. The history of the universe and the elements it contains, in broad strokes. How did the universe begin? How did stars and planets come to be? What is the nature of the elements and atoms that make up everything around us? How did the Earth form and become suitable to harbour life? This will cover the astrophysical, astrochemical, and geological properties of our study.

The first chapter is mainly astrophysical-based, with an emphasis on star formation, nebulae, and the special properties of planetary systems. We will also attempt to understand just how far apart planetary systems and galaxies are from one another.

The second chapter is where we dissect the chemist's mind, and see what it has to offer. We shall see that stars, black holes, the air we breathe, the ground we stand on, and every living organism, are all made from the same 118 elements, just in different ratios and arrangements. These 118 elements provide the basis for everything that has existed, that exists now, and that ever will exist, as they are constantly manufactured and regurgitated by the stars, in a never ending cycle.

Exploring the Essence of Everything

Once we have gained an understanding of the formations of stars and elements, we will move slightly closer to home. The only planet we have ever truly known, our very own 'Pale Blue Dot'. Its history, how it was formed from the primordial rock of the solar system, and the 4.5 billion years of Earth-based evolution that has led us to the present.

The second section, Vol.2, deals with the knowledge of life as we know it. Once again, we will pose challenging questions, such as: What is life itself? How do we gauge the basis of consciousness? How is all life on Earth connected? How and why does evolution occur? These are just a handful of the questions in this section, as we conduct experiments to simulate early Earth conditions, check for evidence of life in deep-sea vents and impact craters, and see how evolution rules every single fibre of our being.

After our pit stop on Earth, we shall turn our gaze to the heavens once again, but this time with the help of satellites, telescopes, and spaceships. The third section, Vol.3, talks of exoplanets, extraterrestrials, and the search for life. We will channel our primitive hunter gatherer instinct, but instead of food and supplies, we will hunt exoplanets, and gather mountains of data. The methods used by astronomers to detect exoplanets, the types of planets there are, and the ability to communicate directly with the aliens.

As usual, questions will be asked and answered. What do we look for in a habitable planet? How do we know if a planet has signs of life or not? What are the key factors that life requires to thrive? Can we detect alien technosignatures? How would we even conceive of travelling to these planets, in far away parts of the galaxy?

Perhaps we don't have to go so far. Start small and build our way outwards. We will discuss the possibility of life and living on the planets and moons of our own solar system. Can we terraform Mars to be a flourishing habitat in 100 years? Is there really life on the moons of Saturn and Jupiter? And can we use these celestial bodies as a platform to potentially launch ourselves into the great unknown?

The final section, Vol.4, is the most philosophical and futuristic yet. Instead of talking about a couple centuries, or even a few thousand years

Prologue

into the future, we will debate about the future of life spanning the next few million years. But, if we can't even predict what happens tomorrow, how can we predict what will happen in a million years?

This is where educated guesses and the pessimistic/optimistic view comes into play. The Drake Equation, Fermi's Paradox, and Simulation Theory are just a few topics that will be discussed over the course of the last three chapters.

The three types of civilizations are particularly interesting, and will actually be presented in a story format, akin to a sci-fi novel. I will introduce characters across three different timelines in the same human civilization to show how the use and efficiency of energy can progress over millions of years, to eventually colonise not just planets, but the entire universe.

The final chapter will discuss the future of life in the universe. How we could travel to the stars, what will become of humans, how we can potentially preserve or extend life. The final questions of the book are fairly metaphysical, with concepts of AI, genetic preservation, micro dimensional mastery, and the end of the universe, all to be debated.

With all that being said, let us take our first step into the unknown, travelling back 13.8 billion years, to genesis. And a very, very, Big Bang.

EXPLORING THE ESSENCE OF EVERYTHING

1

Stars, Planets, and Everything in Between

"If you wish to bake an apple pie from scratch, you must first create the Universe"

Carl Sagan

Genesis

The first half of this chapter deals with the Big Bang theory, cosmic inflation and expansion, the cosmic microwave background (CMB), the interstellar medium, galaxy formation, and the age of the universe. Since these are mainly astrophysical concepts, I will only focus on the sections that apply to us, in our study of life, its constituent atoms, and the future of the universe. The majority of these topics will not be referenced until the very last section in Vol.4, so think of this as a prep class for future reference. If you are interested in acquiring a deeper understanding in these first few topics, I will provide a selection of book recommendations on page 271.

The Big Bang theory was first proposed by the Roman Catholic priest and physicist Georges Lemaître, in 1927. However, the term 'Big Bang' was only coined in March 1949, by English astronomer Fred Hoyle. He said on a BBC radio broadcast; *"These theories were based on the hypothesis that all the matter in the universe was created in one big bang, at a particular time in the remote past"*.

Lemaître's hypothesis was proven correct a couple years later by the astronomers Vesto Slipher and Edwin Hubble. Slipher and Hubble made measurements of the doppler shift of distant spiral nebulae and found out that these nebulae were moving away, relative to the Earth, causing the light emitted from them to be redshifted.

The concept of doppler shifts are easy enough to understand with the help of an analogy. Think of an ambulance going past you at a relatively high speed. The further away it gets, not only is it accelerating, but the sound emitted also becomes fainter. This occurs because the sound waves emitted from the ambulance are being stretched by the time they reach your ears. This same concept is applied to light, in terms of redshift (movement away as lightwaves, or photons, are stretched towards the red end of the spectrum), and blueshift (movement towards as lightwaves, or photons, are constricted towards the blue end of the spectrum).

Stars, Planets, and Everything in Between

OBJECT RECEDING:
LONG RED WAVES

OBJECT APPROACHING:
SHORT BLUE WAVES

Anyways, once it was accepted that distant nebulae were receding, the next step was to figure out how and why. This proved Lemaître's prediction that this recession was due to the expansion of the universe. He further suggested that if this expansion was extrapolated and projected back in time, it would mean that the universe must have been smaller in the past.

This assumption carries all the way back to a point where the universe would have been nothing but a dense, hot, concentration of mass. A 'singularity' where the fabric of spacetime first came into existence. It is at this super dense state, where we start our journey of the universe.

The earliest phases of the Big Bang are subject to much debate and speculation, as observational data and theory can only go so far back in time. In fact, the time after the Big Bang is broken down into absurdly small margins and periods of time.

For example, the four fundamental forces - the electromagnetic force, the strong force, the weak force, and the gravitational force (we will focus on each of these forces in the next chapter) are hypothesised to have been unified as one single force, at 10^{-43} seconds into the Big Bang. That is 10, preceded by 43 zeros. The unification of the four forces is known as the Grand Unification Theory (GUT). The temperature of the universe at this exact time was around 10^{32} °C, with the Planck length of 1.6×10^{-35} m.

The very concept of a particle breaks down in these conditions, and is probably the most debated topic in astrophysics and nuclear physics at the moment. This theory of quantum gravity will reconcile the quantum theory of matter (the small), with Einstein's theory of general relativity (the big).

This theory could very well change the future of humanity in terms of technology, space exploration, and extending the lifespan of our civilization by millions and billions of years. More on that later.

Back to our Big Bang process, it is now 10^{-37} seconds in. This is where cosmic inflation kicks in. This process caused the universe to expand exponentially, faster than the speed of light, causing temperatures to drop by a factor of 100,000. This cosmic inflation is the reason why our universe has a diameter of 93 billion light years, after only 13.8 billion years of existence. This is a very important concept to remember when we debate the development and future of the universe, in terms of galaxy formation, the shape of the universe, and the production of the first particles.

Over the next few fractions of a second (10^{-36} to 10^{-32} to be precise), a couple things occur. Firstly, the strong force separates from this ball of unified forces, causing the first subatomic particles to bind and hold together. More on this in the next chapter. Next, inflation stops as abruptly as it started, having increased the size of the universe by a factor of 10^{78}.

As the universe had expanded so much, its temperature continued to drop, as well as decrease in density since the same amount of mass had now been spread out over a much larger volume. Soon after this, the weak force and electromagnetic force separated, each force now existing as its own entity, already influencing the universe.

We will stop our discussion of the Big Bang just after cosmic inflation ends. At this point, the universe is filled with a hot gluon-quark plasma between 10^{-10} and 10^{-4} seconds after the Big Bang. We will continue from this exact point in the next chapter, when we discuss how the particles, and the first elements, came into existence. For now, we will continue with the formation of galaxies, stars, and planets.

I mentioned that the Big Bang theory came into play with the discovery of receding nebulae and the expansion of the universe. However, there is another piece of evidence that fits this theory, and was discovered much later in 1965, although predictions had already been made in 1948.

For the next step in the history of the universe, we turn to First Light, and the Bell Telephone Laboratories.

Let There Be Light

Bell Labs has had a storied and prominent role in the research and development of various fields of physics. Co-founded in the 1880s by Alexander Graham Bell, inventor of the telephone, Bell Labs has produced many alumni of note, and has won 10 separate Nobel Prize awards, 5 Turing awards, and even has an Academy Award.

Mainly developed as an electronics and telecommunications company, Bell Labs has given scholars the freedom to carry out their research, conduct experiments, and even change the course of history. The transistor, laser, photovoltaic cell, and several programming languages were invented by physicists and scientists at Bell Labs. For our study of the universe, we will focus on their pioneering efforts in radio astronomy, crucial not only in detecting far away signals, but also for broadcasting signals and messages of our own.

Our story takes us to 1964, in the city of New Jersey. Two physicists working at Bell Labs have been tasked with conducting satellite communication experiments.

Arno Penzias, a German-born physicist, joined Bell Labs in 1958, having completed a PhD on using microwaves to amplify and measure radio signals from the spaces between galaxies. His counterpart Robert Wilson came to Bell Labs in 1962, having also done extensive research on the use of microwaves to amplify radio signals in space. Clearly, these two men knew their stuff on electromagnetic radiation in space, especially that in the microwave and radio wave frequency.

Penzias and Wilson used the Holmdel Horn Antenna, a massive horn-shaped antenna measuring 15 metres in length, with a radiating aperture of 6.1 metres by 6.1 metres. The antenna was constructed in 1959, and was at the time one of the largest radio antennas ever built. Sadly, it hasn't been used for several decades, but it is a designated National Historic Landmark due to the work done by Penzias and Wilson. A picture of the antenna has been attached in the gallery at the end of this chapter.

When Penzias and Wilson started to use the antenna, they noticed a sort of background noise (like static in a radio) coming from all directions, in the microwave range. Everyone assumed that this 'extra radiation' came from the surrounding city, or a group of surrounding pigeons that had nested in the horn of the antenna, or even the antenna itself. However, after a few seasons of research, the noise remained and it was concluded that it couldn't be originating from any Earth based source, or even from any extraterrestrial sources in our galaxy, due to the absolute evenness and magnitude of the reading. It had to be coming from outside our galaxy.

At the same time, Robert Dicke and a few colleagues at Princeton University had been discussing theories about the Big Bang. At the time, there were two popular theories on the origins of the universe.

The steady state model stated that the universe was without a beginning or an end, and is forever unchanged in its density, due to a steady creation of matter. The other theory was the Big Bang theory. In the present day, the steady state model has been debunked due to observational evidence pointing towards a Big Bang theory, but the steady state model was all the rage back in the day.

As previously stated, the Big Bang theory only came into play after Slipher and Hubble's discovery of receding nebulae, proving Lemaître's hypothesis. However, due to a lack of evidence, half the physicists of the time found it hard to believe. They needed something more concrete and more importantly, observable. This is where our Cosmic Microwave Background enters the story, with the groundbreaking discovery by Penzias and Wilson.

Robert Dicke and his colleagues elaborated on the existing Big Bang theory by saying that if there had been such a large explosion, there should also be residue from said explosion. This isn't fallout in terms of shrapnel from a bomb, although the concept is similar, this residue takes shape in the form of light and radiation.

The reason that the signal detected was in the microwave end of the electromagnetic spectrum, is due to the amount of time that had passed, and the distance the wave had travelled since its creation. Initially, it would have been a sort of high-energy, high frequency, gamma ray, but it had since been stretched and redshifted to a microwave. If the universe continues to expand, the Cosmic Microwave Background will become the Cosmic Radio Wave Background in around 100 billion years time. A sample of the electromagnetic spectrum is attached for reference.

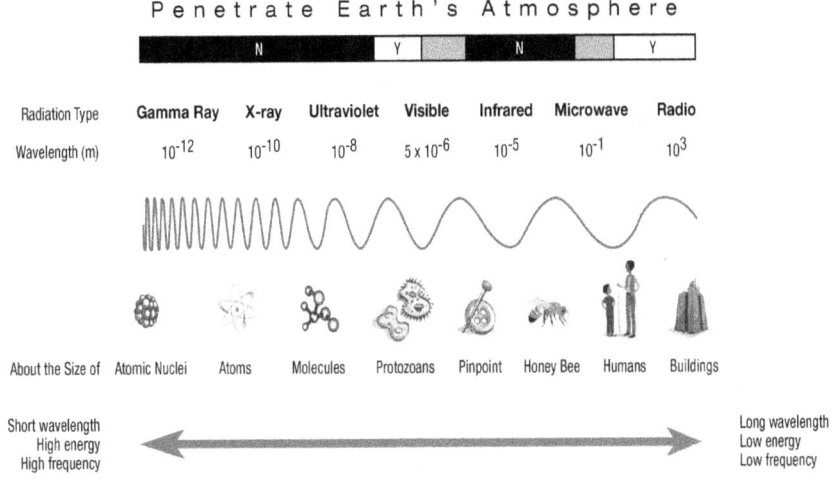

One of the most foolproof ways of confirming the Big Bang model was to estimate the current temperature of the universe, which has been cooling down ever since the expansion of space and cosmic inflation began.

Many scientists made many predictions over the years. In 1948, Ralph Alpher and Robert Herman estimated the temperature of the universe to be around 5 Kelvin. A few years later, George Gamow estimated it to be around 6-7 Kelvin. Robert Dicke and his gang estimated the temperature to be 3 Kelvin. This means that there should be constant radiation from space, in the microwave region, with a temperature of 3 Kelvin. What did Penzias and Wilson detect in 1964? Constant excess radiation. What form did it have? Microwaves. What was its temperature? 3 Kelvin. Three strikes and you're out, steady state model.

This discovery all but sealed the Big Bang model as the leading theory for the origin of the universe, and Penzias and Wilson won the Nobel Prize in 1978 for their work in CMB and radio astronomy.

But what is the Cosmic Microwave Background? Leftover radiation? Cooled down temperature? First light? What can it be defined as exactly, and why is it important to us?

In our universe, we have matter, and we have radiation. Matter exists in the form of atoms, consisting of protons, neutrons, and electrons, whereas radiation exists in the form of photons. However, there was once a time when matter and radiation were pretty much one and the same. For a period of 380,000 years after the Big Bang, the universe was so hot that electrons couldn't bind with protons to form atoms, leaving them suspended in space. As a result of this 'sea of electrons' the photons of the time constantly collided with them and were unable to spread out. (This same process occurs in our Sun, with photons eventually leaving some 100,000 years after they are first formed in the core). In essence, there were photons but no light, since the photons were trapped in this super hot plasma, unable to travel throughout the universe. This is known as the opaque era of the universe.

Shortly after this time, the universe had cooled down enough to allow the electrons to combine with protons to form the first hydrogen atoms, allowing the photons to finally break free and scatter throughout space. This is the so-called 'first light' of the universe. The initial scattering of photons (radiation), 380,000 years after the Big Bang.

It is this very radiation which Penzias and Wilson detected in 1964. Radiation that had been travelling for 13.8 billion years to reach the Horn Antenna. Radiation that had cooled down from 3000 Kelvin to 3 Kelvin. Radiation that had gone down the electromagnetic spectrum, from gamma rays to microwaves. That is the Cosmic Microwave Background. It is all around us, at every second, in every place. The earliest snapshot of a primordial hot universe. We have detected photons that have been travelling at the speed of light, for 13.8 billion years, penetrating every nook and cranny of spacetime. We have taken a picture that is essentially 13.8 billion years old. How is that for perspective?

The CMB has since helped us to calculate the age of the universe (13.8 billion years) and the temperature of space (from 3 Kelvin, it has now been fine-tuned to 2.7 Kelvin [-270.45 Degrees Celsius]) it provides the best data we have on the early universe, and is still being researched using more precise telescopes and instruments to travel further back in time. Tiny fluctuations in temperature and the evenness of the CMB have proven the theory of cosmic inflation, and have also provided insight on the differences in density that formed the first galaxies.

The next stop in our trek through time puts us in an uncertain phase. It is 380,000 years after creation and the universe has cooled down enough for hydrogen atoms to form, but how do we go from a tiny atom to massive galactic structures? And where are we, relative to other galaxies? Are we some stray unique galaxy, outcast from its peers? Are we closer to other galaxies and display some sort of commonness? Let us now discover the intricate details of galaxy formation and learn of our place in the universe.

A Hitchhiker's Guide

In the previous two sections, I have mentioned cosmic inflation and tiny fluctuations or 'unevenness' in the CMB. It is actually this unevenness, paired with cosmic inflation, which is responsible for the structure of our universe today. Why are galaxies so far apart from each other? Wouldn't it make more sense to have them closer together? It's like building two tiny towns, on opposite sides of a massive planet and providing them no means to travel from one place to the other. But perhaps we should be thankful for these fluctuations in the fabric of spacetime, for giving us something, instead of nothing.

Exploring the Essence of Everything

We will tackle two concepts in this section. First, we will learn how tiny fluctuations in space time are responsible for the formation of galaxies. Then, we will learn of our place in the universe, and how far we are from everyone else. This second part will be especially vital when we debate the logistics of travelling to other stars and planets, as well as the distance and time threshold imposed on us in our search for alien life.

Before inflation, there were tiny quantum fluctuations in the fabric of spacetime. What does this mean? A quantum fluctuation is basically the random change in the amount of energy in a certain point in space. So sometimes it would have more energy, sometimes it would have less. According to quantum field theory, these fluctuations could produce matter. The main rule about quantum fluctuations that we have to remember is that they are completely random and unpredictable.

So these fluctuations existed before inflation. Regions of higher and lower energy. During inflation, these fluctuations were amplified and spread out, but remember, they didn't all have the same energy level upon spreading out. So now we have more matter in higher energy areas, and less matter in lower energy areas. This creates differences in density.

Now we have a universe that is still hot, with areas of higher and lower density, but no atoms yet. Remember, atoms didn't exist until 380,000 years after creation. Once the universe had cooled down enough to allow protons and electrons to bind and form hydrogen atoms, two things occurred that are observable today.

The first is from the CMB. Remember how I mentioned that the CMB has tiny differences in temperature? That there are some regions that are slightly hotter, and some regions that are slightly colder? This is due to those differences in density. A more dense area of space would have had more photon-electron mass, and thus, more photons would have been released from that area of space. More photons being released would have caused that region to have a higher temperature and energy level. Vice versa for less dense areas of space. This caused the tiny fluctuations in temperature in the CMB.

Stars, Planets, and Everything in Between

The second thing that occurred, is that gravity began to take effect. To understand how gravity influenced these first atoms, we can use a simple enough analogy.

Let's say, we have a volume of space (volume A) with 100 hydrogen atoms of the same mass. On the other side, we have another volume of space (volume B), with the same fixed volume, but with only 50 hydrogen atoms. Volume A has a higher density since it has double the number of hydrogen atoms and hence, mass. Differences in density can be caused by differences in mass, volume, or both. For this example, we will use a difference in mass.

Before we continue, let us understand some basics about gravity. Gravitational force is the amount of attractive force that two objects exert on one another, by warping the fabric of spacetime. The rule of gravity obeys an inverse square law of distance to gravitational force. Essentially, the closer the distance between two objects, the stronger the force of gravity between them, and the force of attraction will be much stronger. Also, if there is more mass in a certain region, it will be easier for those atoms to attract each other, than if there were less mass, since spacetime will be more warped. Gravitational force increases as mass increases.

Going back to our two volumes, since there are more atoms in volume A, it will be easier for them to attract each other, and move closer together. As the atoms accumulate together, the region becomes more and more dense and attracts atoms that are even further away. For volume B, this is not the case. Because there are less atoms, and the lower mass is spread out over the same amount of volume, it is harder to attract other atoms, and this region remains unchanged in its density.

Over time, volume A becomes even more dense, and volume B becomes less dense, as it loses its atoms to areas that are now denser. It is this primary difference in density that causes galaxies to form, as more and more matter is accumulated in the denser regions, and the space between the dense regions remains undeveloped and unchanged. That is why we have to thank the quantum fluctuations for creating these regions of differing densities, which eventually led to the formation of the galaxies and everything in them.

Now that we know how the galaxies are formed, what of the distances between them? How long, on average, would it take to travel from one galaxy to another? Is it even worth the effort to do so? Perhaps we need to decrease our scale and observe star systems in our own galaxy instead. We shall see the position of the Earth in our own galaxy (the Milky Way), and the position of the Milky Way relative to the scale of the universe.

Most galaxies are divided into four categories. Elliptical, lenticular, spiral, and irregular. Galaxy types differ based on their composition, shape, age, and size.

Our galaxy, the Milky Way, is a subsection of the spiral type, called a barred-spiral galaxy. This is because the central bulge looks elongated, like that of a bar, with the spirals emerging from either end. The Milky Way is about 100,000 light years across (a light year is the distance travelled by light in one year, moving at a speed of 3×10^8 m/s), and contains around 100 billion stars.

The Milky Way actually derives its name from the Greek myth about the goddess Hera who sprayed milk across the sky. In modern culture, it gets its name due to the milky white appearance of the band of light in the sky, when viewed from a dark area.

The majority of galaxies contain a supermassive blackhole in their centre, around which all the stars, planets, and celestial objects orbit. Our own Sun orbits the galactic centre, and takes 250 million years to complete one single orbit. This is called a cosmic year, and in reference to our Cosmic Calendar, it would occur once every 6.75 days. The supermassive blackhole in our galactic centre is called Sagittarius A*, and it has a radius of 12 million kilometres. For reference, the radius of the Earth is just 6,371 kilometres. It also has a mass of 4.3 million times the mass of the Sun. It was this exact black hole which garnered much publicity in 2022, when an image of it was released to the public. This image was taken in 2017, but took 5 years of calculations to process.

Stars, Planets, and Everything in Between

To put our place in the Milky Way into perspective, we will work from the Earth outwards. We are on Earth, which orbits the Sun every $365\frac{1}{4}$ days. The Sun is just one star in our stellar neighbourhood of around 50 stars, covering a radius of 10 light years. The closest star to us is Proxima Centauri, just 4.2 light years away. A hop, skip, and jump away in relation to universal scales. Our stellar neighbourhood is actually a fairly quiet area in the Milky Way.

If the galactic centre was a bustling city, we would be in the suburbs, on the outskirts. The solar system is located at a radius of 27,000 light years away from the galactic centre, about halfway from the edge to the centre.

So before we think of colonising other galaxies, we should focus on reaching the star systems nearby, possibly Alpha Centauri, Epsilon Indi, or Gliese 65. We will continue this specific topic in volumes 3 and 4, when we discuss interstellar travel.

For conversation's sake, let us travel out of the Milky Way and see

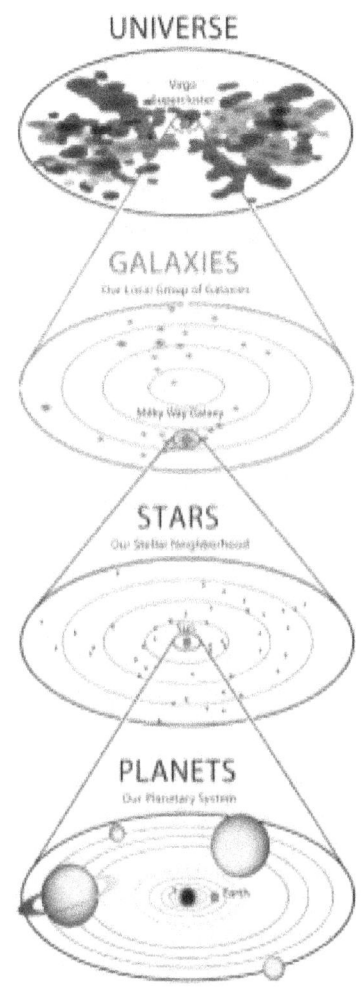

how far away we are from other galaxies. How far did inflation push apart those primordial quantum fluctuations?

The Milky Way is part of the Local Group, which it dominates along with the Andromeda galaxy, our closest neighbouring galaxy. The Local Group has a total diameter of roughly 10 million light years, and contains a mixture of elliptical, spiral, irregular, and dwarf galaxies. The Andromeda galaxy is about 2.5 million light years away from us. This means that even if humans ever achieved space travel at light speed, it would still take us 2.5 million years to reach the closest galaxy. For reference on our Cosmic Calendar, 2.5 million years ago would have been on the 31st of December, 10:24 pm. This means that the entire history of human civilization and culture could have occurred, in the time it would take us to reach the nearest galaxy.

The Local Group is part of the Virgo Supercluster, containing at least 100 galaxy groups and clusters within its diameter of 110 million light years. The Virgo Supercluster is one of about 10 million superclusters in the observable universe. Feel small yet?

The Virgo SC is one quarter of the Laniakea Supercluster, home to around 100,000 galaxies. It has a diameter of around 520,000 million light years, and a mass 100,000 times the mass of our galaxy.

The Laniakea SC is part of the Pisces-Cetus Supercluster Complex, which is a galaxy filament. Galaxy filaments are the largest known structures in the universe, consisting of walls of galactic superclusters. The Pisces-Cetus Complex is estimated to be 1 billion light years long, and 150 million light years wide. Our massive Virgo SC accounts for just 0.1% of the mass of this massive galaxy filament. The best part? This galaxy flament is ten times smaller than the largest galaxy filament in the universe. The Hercules-Corona Borealis Great Wall is the largest structure in the universe, with a diameter of 10 billion light years. Remember what the total size of the observable universe is? 93 billion light years in diameter. We are almost obsolete in the grand scheme of things.

We have talked at length now about these galaxies and how they were formed. But what of all the empty space between the galaxies and stars? It seems like an awful waste of space, with no potential for vacancy. Perhaps it isn't as void as first thought. We will now venture into the great nothingness between everything.

The Great Abyss

In our universe, we have three types of 'empty space'. There is the space between the planets in a star system, named interplanetary space, there is the space between the stars in a galaxy, named interstellar space, and there is the space between the galaxies themselves, named intergalactic space. Interplanetary for planets, interstellar for stars, and intergalactic for galaxies. Simple enough. We will focus on these three voids in this section.

Before we get into the details of each space, I'll first demonstrate the differences in density and matter between these three regions, using our analogy from the previous subsection. Let us go back to our two volumes, volume A (more dense), and volume B (less dense).

Remember, volume A formed the galaxies, and volume B formed the space between the galaxies. Interplanetary and interstellar space formed from volume A, while intergalactic space is volume B. Interplanetary space is the space that fills a star system, and has a density of between 5-40 atoms per cubic centimetre. This density varies since the planets are much closer together than the stars and galaxies are.

Interstellar space is the region between stars in the same galaxy, and saying that, it has a density of one atom per cubic centimetre. It is essentially a less dense region, inside the already dense region of volume A. So we see that even inside the galaxies themselves, we have areas of higher and lower density, due to the force of gravity once again.

Intergalactic space on the other hand, volume B, has a density of one atom per cubic metre. This means that interstellar space is 100,000 times denser than intergalactic space. This makes sense, considering that interstellar space comes from the denser region of volume A, and intergalactic space is the less dense region of volume B. With this knowledge, let us now look at each region in more detail.

Interplanetary space and its constituents are mainly influenced by the star in its centre. The star's outer layer, the corona, ejects a stream of charged particles called solar wind. This solar wind causes dust, cosmic rays, and hot plasma to be present in the interplanetary medium. This is also the only region out of the three to have other force fields in it. Gravitational, magnetic, and electric fields permeate the region between the planets and the central star.

It also has the most varying temperature, due to the direct influence of its star. The average temperature of solar winds is around 1 million Kelvin, while the dust particles near Earth have an average temperature of around 283 Kelvin. Clearly a large difference here. This is once again down to the rule of inverse square with distance, as it is in gravity. Essentially, the further away the distance from these solar winds, the faster the temperature gets cooler, as it decreases by a squared factor. For dust particles in the asteroid belt, we can expect temperatures between 165 to 200 Kelvin. Keep in mind, that these are just the temperatures of the particles and winds present in this space, not the space itself. Moving further out from the Sun, the surface temperature of the planets gets considerably colder, with the Earth in the 'habitable zone', the distance from the star where surface temperatures enable water to exist as a liquid.

The interaction between this space and planets, depends on the presence of magnetic fields. Bodies with little to no magnetic field are directly impacted by solar winds and radiation. Planets like the Earth and Jupiter with an appreciable magnetic field disrupt the flow of the charged particles in the solar wind, which are then channelled around it, towards the poles. These charged particles can leak into the upper atmosphere, where they collide with atoms and molecules, causing the aurorae over the poles.

The topics on heat distribution throughout interplanetary space, the habitable zone, and the effect of solar wind on planets will be examined in more detail when we learn about the habitability of exoplanets.

Stars, Planets, and Everything in Between

Let us move out of the hustle and bustle of interplanetary space, to the quieter suburbs of interstellar space. The interstellar medium is the matter and radiation that exists between stars in a galaxy. Coincidentally enough, the constituents are similar to that of interplanetary space. Cosmic rays, dust, and gases (this time in atomic, molecular, and ionic form).

The composition of the interstellar medium is 91% hydrogen, 8.9% helium, and around 0.1% heavier elements (any elements heavier than hydrogen and helium are known as 'metals' in the astrophysical circle, but I will refrain from using this terminology for the sake of simplicity). These percentages will be important to remember in the next chapter, when we discuss why there is so much more hydrogen compared to other elements.

The main celestial objects in interstellar space that we are really interested in are nebulae. As with everything else, interstellar space has regions of higher and lower density. Nebulae are regions of higher density, filled with ionised, neutral, or molecular hydrogen, helium (and sometimes heavier elements), and cosmic dust. Nebulae are essentially regions of star birth, and the effects of star death.

There are two ways in which a nebula can form; by the condensation and cooling of gas already in the interstellar medium, or from the material shed by a star in the latter stages of its stellar evolution. The shedding of material from a dying star depends on the star's mass. The material is either lost from a planetary nebula, or from a supernova explosion.

There are many types of nebulae. They are mainly classified based on their composition, age, size, process of formation, and method of detection. The three main types are; diffuse nebulae, planetary nebulae, and supernova remnants. Examples of each of these types of nebulae are provided in the gallery of this chapter.

Diffuse nebulae are the most common and broadest type of nebulae. They are named thus as they are extended and contain no well-defined boundaries. They can be broken down further into three different variations; emission nebulae, reflection nebulae, and dark nebulae.

Emission nebulae are also known as 'stellar nurseries' and are the most common form of nebulae. In these star-forming regions, gas, dust, and other materials cluster together to form denser regions. The density attracts more matter and eventually becomes dense and hot enough to cause the formation of stars. The remaining material is then used to form planets and other objects such as asteroids, moons, and comets. Emission nebulae are largely composed of ionised hydrogen which emits radiation, hence the name. The most active and closest emission nebula to Earth is the Orion Nebula, which is visible with the naked eye, just south of Orion's belt in the constellation Orion.

The second type of diffuse nebulae, the reflection nebulae, are clouds of interstellar dust which reflect the light of nearby stars. The energy from these stars is insufficient to ionise the gas to form emission nebulae, but is enough to provide sufficient scattering to make the nebulae visible. The frequency at which these nebulae are detected is similar to that of illuminating stars. The principle of scattering can be observed in our red horizon and blue sky. This is mainly because the scattering of blue light is more efficient than red, giving these nebulae a distinctly bluer hue.

The last type of diffuse nebulae are dark nebulae. These are also referred to as absorption nebulae, due to their ability to absorb and block visible light from background objects such as stars or other types of nebulae. The extinction of light is mainly caused by interstellar dust grains located in the coldest, densest part of molecular clouds in the nebulae.

The second main type of nebulae are the planetary nebulae, which are formed when a mid-mass star sheds its outer gaseous layers after nuclear fusion in its core has ceased. As the dying star loses material, its temperature rises. The ultraviolet rays emitted from this increase in temperature causes the gaseous layers emitted to become ionised and charged, leading to more star formation.

Generally, planetary nebulae contain heavier elements (carbon, nitrogen, oxygen) than diffuse nebulae due to nuclear fusion. These elements enrich the interstellar medium and the subsequent generations of stars that are formed from these planetary nebulae will have higher metallicity (the abundance of elements heavier than hydrogen and helium). Our own Sun will become a planetary nebula in around 5 billion years.

Let's hope that by then we would have mastered the ability of space travel and planet colonisation.

The third and final main type of nebulae are the supernova remnants. As the death of a mid-mass star results in a planetary nebula, the death of a high-mass star results in a violent supernova explosion. The expanding shell of gas from the explosion forms a supernova remnant. The heat from this explosion causes the gas to become ionised, and allows for more star formation. The stars formed from supernova remnants have the highest amount of metallicity, due to the much higher abundance of heavy elements formed in the explosion. The detailed process of how these heavier elements are formed will be covered in the next chapter.

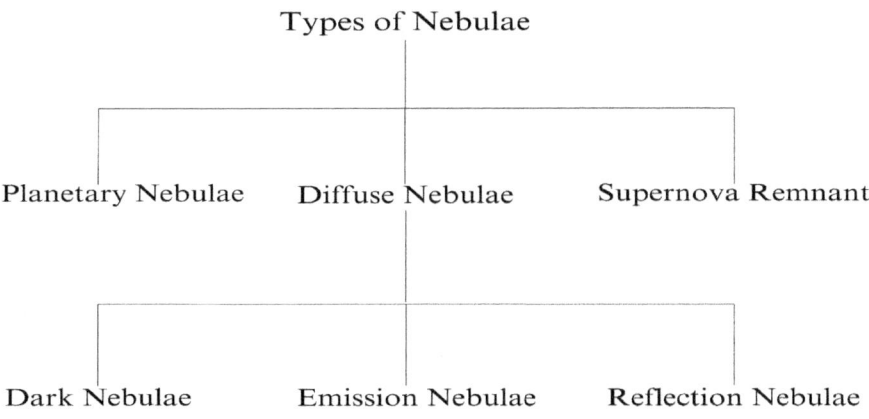

Alas, we must leave these wonderful, polychromatic nebulae behind, with their myriad of colours and stars. The last and largest type of space in our universe is intergalactic space, the space between galaxies. These voids of space are roughly 21 million to 90 million light years long, and are thus, one of the lowest density regions in space. However, this doesn't mean that the gas in these regions is any less special and beneficial. In fact, due to the sheer size of these regions, they actually contain more matter than the other two previous spaces put together.

Intergalactic space is permeated by a rarefied plasma, which is known as the intergalactic medium, and is mostly made up of ionised hydrogen. This supercharged plasma emits energy in the x-ray spectrum due to the ionisation of hydrogen atoms, and has been detected by NASA's Chandra X-ray Observatory.

These three 'empty spaces' in the universe are actually connected quite systematically. The gas from the intergalactic medium will flow into galaxies over millions of years, replenishing the star forming nebulae in the interstellar medium. In turn, the interstellar medium forms the star systems which contain the interplanetary space. Evidently, these spaces between everything have actually been the building blocks for everything that has been, everything that is, and everything that will be.

Now that we know what is in between the stars and galaxies, let's take a closer look at the celestial objects inside the star systems and galaxies themselves. From neutron stars and black holes, to suns and planets, this is the tale of how all the matter in our universe takes shape.

Of Everything Spherical, or Otherwise

If you've ever watched a video of astronauts aboard the International Space Station, you have probably noticed a phenomenon that is also observable on Earth. When astronauts pour water in the ISS, the water immediately forms spherical droplets suspended in mid-air. Now, if we pour water on Earth, it falls in a steady stream due to the effects of gravity. However, if you observe dew on a plant, the water droplets are spherical in shape. Why does the concept of spherical droplets apply to some situations but not others?

The next time you wash the dishes or take a shower, observe the shape of the soap bubbles. Perfect spheres. What of the water poured aboard the ISS? Perfect spheres. Ok, well what about the shapes of stars, planets, and moons across the universe? Perfect spheres. Why is this so? Why were spheres chosen as the geometrical shape of choice across the universe? Why not cubes or cylinders or pyramids? Why do the forces of nature seemingly force a spherical mould on all naturally occurring objects? There seems to be a hidden thread between all these spherical objects.

Two forces are at work here. Surface tension and gravity. We have discussed gravity briefly, so let's turn our attention to surface tension for the moment. Surface tension is essentially the tendency of liquid surfaces at rest to shrink into the minimum surface area possible due to the attractive forces between their molecules. The molecules in the liquid will try their best to get as close together as possible. This shrinking has to occur at a uniform and even rate, meaning that all surfaces have to shrink simultaneously. A bit of high school calculus can tell us that the smallest possible surface area to volume ratio occurs in spheres. Thus, liquids will form spherical shapes due to surface tension. This effect is more pronounced in space due to the lack of gravity as the water molecules are only influenced by their own cohesion forces and surface tension.

Alright, this explains the spherical shape of liquids, but clearly surface tension doesn't work with solids. So why do we see naturally occurring solids in spherical shapes as well? Force number two; gravity.

The concept of gravity on spheres is probably easier to understand than surface tension (of which I have excluded the majority of the theories and equations, and have just provided the board strokes). We know that gravity decreases or increases based on an inverse square law. We also know that gravity attracts all objects inwards towards its centre of mass. Finally, we know that the gravitational force of an object must be as equal as possible throughout its volume. Imagine if we had some areas of Earth at 1 G, and other areas at 10 Gs. The key principle that reconciles all of these rules is that gravity is measured from the centre of an astronomical body, not its surface. This means that the distance from the centre of the object to its surface must be equal in all directions. I'm sure that most of you will be familiar with the concept of radius. What is the only shape that is equidistant from its centre to its edge no matter the direction? A circle. What is a circle in 3 dimensions? A sphere.

There is one main exception to this rule. As mass and density decreases, the gravitational force of an object decreases. If the gravitational force of a celestial object decreases enough, the gravity pulling everything together in a firm sphere will be overcome by the forces of the chemical bonds between the molecules and rocks in the object. This is mainly evident in smaller, lower mass objects, such as comets, asteroids, and the potato-shaped moons of Mars, named Phobos and Deimos. This is also what causes mountain ranges to form on the surface of planets, as the force of gravity decreases away from the centre. However, the height of these mountains compared to the radius of the planet is so small, that the planet is taken as a smooth sphere anyways.

The final piece of our spherical jigsaw is the effect of centrifugal force on the perfect shape of spheres. When objects rotate or spin, centrifugal force acts outwards from the edge of the object, balancing the inward force of gravity. As the object spins faster, the centrifugal force increases, preventing a total collapse due to gravity. We can apply this theory to the formation and development of planets and galaxies.

The Earth is not really a perfect sphere. It is in fact slightly squashed at the poles, and elongated at the equator. Think of a burger. Flat on top, but elongated on its sides. Alright, perhaps the reality is not that drastic, but we see that centrifugal force has caused the Earth to bulge and flatten slightly due to the speed of its rotation. This is not the same as a flat Earth theory! Remember, the main factor in centrifugal force is the speed of angular rotation. Since a planet spins much faster than its star, the planet is slightly deformed, while the star is perfectly spherical.

This can be extrapolated to the shape of some galaxies. If we observe the band of stars across the Milky Way, we notice that they all lie in a relatively straight line. In 3 dimensions, this would be a disk. But didnt I just say that all naturally occurring objects will form spherical shapes? Why the need for disks all of a sudden?

There is another factor which affects centrifugal force, that is directly tied to gravity; mass. Galaxies are one of the heaviest objects in the universe. As a result, they will also have some of the strongest forces of gravity. A stronger force of gravity will need to be balanced by an equally strong centrifugal force to prevent the entire galaxy from collapsing inwards.

The Milky Way certainly started out as a spherical mass of gaseous matter. Over time, this matter spun round and round, increasing in speed. This, along with its large mass, increased the centrifugal force. Thus, the strong centrifugal force caused the galaxy to flatten more and more over billions of years, until it became the disk shape we observe today in the night sky. Note however, that this does not apply to every single type of galaxy.

With this prerequisite knowledge, let us now see how these spherical objects differ in size, composition, density, temperature, and habitability. We can use a scale of decreasing size to show the relationship between the objects in our universe.

Firstly, the largest structures in the universe; galaxies. As previously stated, not all galaxies obey the spherical law of nature, and will differ based on mass, size, and composition. Inside the galaxies, the next biggest objects are the nebulae. As we have already touched on galaxies and nebulae in the previous two subsections, I will not repeat the characteristics of these two objects. A twice told tale is not worth telling.

After nebulae, we move into the stellar systems formed from said nebulae. The objects in these systems can be placed into 3 groups; stars, planets, and simple bodies.

A star is the largest, heaviest, and densest object in a stellar system. It is formed from the bulk of the molecular cloud or nebula. The formation process and life cycle of a star will be discussed in detail in the next subsection, for now we will focus on its main characteristics. The star consists of 90% hydrogen, 9% helium, and the remainder existing as heavier elements, in terms of number of atoms. In terms of mass, the star consists of 73% hydrogen, 25% helium, and the remaining mass in heavier elements.

The high temperatures and gravitational force of the star causes nuclear fusion to occur (the entire process of nuclear fusion will be discussed in chapter 2) in the core of the star, producing heavier elements. Obviously

Exploring the Essence of Everything

these high temperatures make habitability on the surface of the star impossible. Finally, the reason a star spins much slower and hence has a much more spherical shape compared to its planets, is due to the turbulence in the star's stellar wind caused by its magnetic field. This causes it to slow down, a form of 'magnetic braking'.

The next biggest objects in a stellar system are the planets. Planets differ in size, from really large ones (Jupiter and Saturn) to rather small ones (Mercury and Mars). Even the composition differs, as the planets get further away from its star. The planets closer in will be made of rock, and the planets further out will be made from mainly gas and ice. However, this isn't a hard and fast rule as we shall see. The density of planets depends on their mass to volume ratio, mainly dependent on their composition, which also has a direct effect on their gravitational force. The temperature of a planet is measured by its amount of thermal emission and the energy of these emissions. Of course, the further away a planet is from its star, the cooler it will be. Finally, the habitability of a planet depends on a plethora of factors. Its temperature, its gravitational force, its pressure, its atmosphere, the type of star that it orbits. The list goes on and on. These factors will be the main debate of chapter 8.

The final objects in a stellar system, and the so called 'dregs' of the molecular cloud that formed the system, are the simple bodies. These consist mainly of moons, asteroids, comets, and meteoroids. As the size and mass of these objects decreases, so does its gravitational force, giving the majority of them asymmetric, skewed shapes.

Moons, or natural satellites, are objects that orbit a planet, and are usually leftover material from the formation of the planet. The bigger the planet, the more matter leftover. There are 297 known moons in the solar system, of which only 19 have enough gravity to be round, Earth's Moon among one of them. The abundance of moons increases as we get further from the Sun since the size of the planets themselves increases. Mercury and Venus do not have any moons, Earth has one, and Mars has two deformed moons, Phobos and Deimos. Jupiter has 95 known moons, of which 57 have been named. Saturn has a whopping 146 moons, however

Stars, Planets, and Everything in Between

most of them are quite small and are formed from the icy rings around Saturn. We will take a closer look at some of the moons of Jupiter and Saturn as potential sites for life in chapter 8. Uranus and Neptune have 27 and 14 moons respectively.

Asteroids can be thought of as small rocks that did not accumulate enough matter to become fully fledged planets. To put into perspective just how small and light asteroids are compared to the planets, the combined mass of all the asteroids in our solar system (of which there are around 1 million), is just 3% of the mass of the Earth's Moon. However, this doesn't mean that they are simply barren wastelands of rock floating around the planets. Asteroids provide immense potential to space travel when we consider the possibility of asteroid mining for rare elements, and space bases to reduce the cost of space travel. Some asteroids have even been hypothesised to contain water in the form of ice, and certain sugar molecules. Potential for life? We shall see.

Comets are small, icy bodies, and are the farthest objects in our solar system. The difference between asteroids and comets is easy to understand. Asteroids are mainly composed of rock and are thus found much closer to the rocky, inner planets. Comets are mainly composed of ice and are hence found further out in the outskirts of the solar system. Comets produce elongated tails of gas and dust due to interactions with the solar wind and radiation. Comets have highly elliptical orbits around the Sun, some taking 200 years to orbit, others taking 20,000 years to orbit. Similar to asteroids, certain amino acids and sugars have been found on the surfaces of comets. Perhaps life in our universe is not so rare after all. Or are we all descendants from asteroids and comets?

Meteoroids are the smallest objects in a stellar system. They consist of anything from the size of atoms to sizes just smaller than that of an asteroid. These particles have either been left over from the formation of the system, are debris from collisions between bigger objects, or fragments from comets and asteroids. Around 25 million meteoroids enter Earth's atmosphere everyday, the majority of which burn up in the atmosphere before they can come into contact with the surface.

With that, we have completed our grand tour of all the heavenly bodies in our universe. From the large galaxies and nebulae, to the planets and moons in our own backyard. We have already learnt a decent amount concerning galaxies and nebulae, so let us devote the remaining subsections of this chapter to the stars and planets.

The Circle of Life

There appears to be a beautiful symmetry in the life cycle of stars. In many ways, it is the ultimate reuse, reduce, and recycle analogy. In fact, the life cycle of stars can be thought of as the greatest loop in the universe; constantly moving between living and dying, and always keeping the next generation in mind.

One usually finds it troublesome to find the starting point in a circle, so I shall start with the birth of stars. For this, we pay a visit to our tried and trusted friend, the nebula. Let us set the scene, and watch as this great story of life and demise spins its threads throughout the universe. You may use the image shown as a visual representation of this cycle. The stage is set. The cast is ready. All the pieces are in play. And away we go.

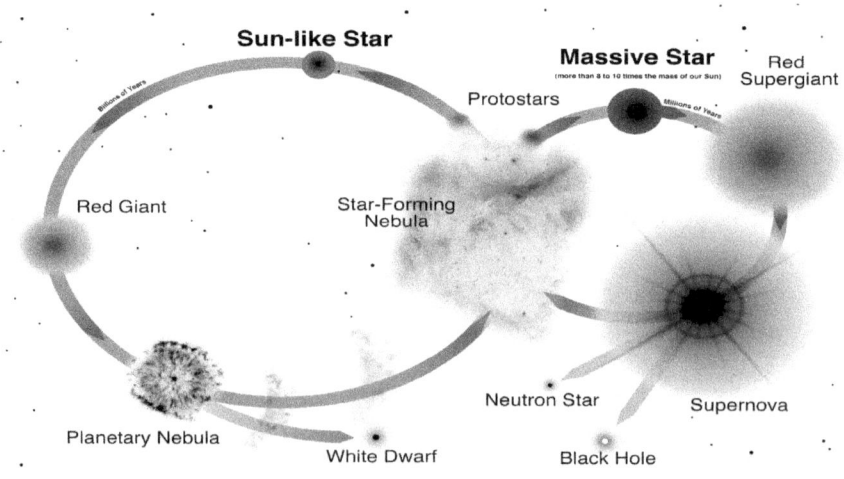

Stars, Planets, and Everything in Between

Act 1 Scene 1

Our story starts in two relatively quiet regions of space. A handful of stars are visible, permeating the absolute darkness like so many grains of salt on a black canvas. It is cold. It is hostile. It is unforgiving. Nature spared no mercy when designing the environments of space.

It is 5 billion years after the Big Bang. This exact period of time does not really have any significance to our story, it is merely a way to provide scale and perspective to the reader. It is May 13th on our Cosmic Calendar. Space has been expanding and pulling the galaxies apart since time immemorial. Already, particles that were together in that primordial soup of protons, electrons, and photons will never meet or interact again, as long as the universe exists.

In the first region of space, in a slightly larger-than-average galaxy, in the space between the stars, there exists the building blocks of everything. We are in the presence of a nebula. To be exact, it is an emission nebula (or stellar nebula). This is the birthplace of stars. In this stellar nebula, there is an average sized molecular cloud, with elements that have been gathered from the surroundings. 90% of the elements are hydrogen, 9% are helium, and the rest are slightly heavier elements. We will name the molecular cloud in this region of space, Aristotle.

50 million light years away from Aristotle, in another far away galaxy, in another equally quiet region of space, we find another stellar nebula. This stellar nebula has a molecular cloud with the exact same composition as Aristotle, but it is much bigger. We will name it Plato.

So our tale starts with these two molecular clouds, Aristotle and Plato. The names do not signify anything, it is just for simplicity's sake. Aristotle is an average sized molecular cloud, Plato is much larger. However, the composition and ratios of hydrogen to helium to heavier elements is the same. With that, let us move on to the next scene.

Exploring the Essence of Everything

Act 1 Scene 2

Thus far, Aristotle and Plato have lived relatively tame lives. Not much has occurred apart from the churning of the gas and molecules inside them. They have been accreting material from the surroundings and are slowly growing in mass and size. Similar to the early universe, there exists regions of higher density and lower density.

In these two molecular clouds, there exists a sort of equilibrium. Two forces directly opposed to each other, just as Newton said. We have a force pushing outwards, and a force pulling inwards.

Needless to say, the force pulling everything in is our good friend, gravity. It keeps the gases in Aristotle and Plato from dispersing throughout the interstellar medium. As always, denser regions will have higher gravity and vice versa.

The force pushing out is the internal pressure from the collisions of the atoms and molecules with each other. When these particles collide, they produce heat and a force which pushes out against gravity, preventing the collapse of the entire structure. Once again, denser regions will be concentrated with more particles, which will cause more collisions, which will create more outwards force to balance out the strong inwards force of gravity. Wherever we look, there appears to be a splendid symmetry and equilibrium permeating our universe.

This beautiful balance stays in place until it reaches a point where the scales are tipped in favour of a force. As the molecular clouds continue to get bigger and heavier, the force of gravity is too strong for the outwards pushing force. When this happens, it triggers a point of no return. Aristotle and Plato are said to have collapsed under their own gravitational force.

Gravity now starts to pull the molecules closer together in a runaway chain reaction. The majority of mass and hence, gravity, will be concentrated in the centre of the cloud, and as is always the case, gets lesser and weaker the further out you travel.

Since Aristotle is an average sized molecular cloud, its gravitational forces are weaker than that of Plato, but is still sufficient enough to collapse and cause this effect where all matter is pulled in towards the centre. And so ends act 1 of our play, of the life and death of stars. The molecular clouds have collapsed, and we move on to act 2.

Act 2 Scene 1

The collapse of Aristotle and Plato causes gas to fall inwards towards the centre of the molecular clouds. This gas is under immensely strong forces of gravity. As always with gravity, this is a runaway effect. The more gas falls in, the heavier the centre gets. The heavier the centre gets, the stronger its gravity. The stronger its gravity, the more gas falls in. And the cycle repeats itself.

As gravity increases, so does the energy level of the particles falling in. This energy is in the form of gravitational potential energy. Energy from this system is lost as heat energy, heating up the gas and surrounding molecules.

For both Aristotle and Plato, this will eventually result in the formation of a protostar. The protostar is formed from the increase in temperature and pressure, and exists as a small rotating ball of superhot gas. The protostar continues to accrete gas and dust, and grows in size.

This is where the sizes of Aristotle and Plato play a big role. Since Plato is much larger than Aristotle, it will form a bigger and heavier protostar. This will cause it to form a much larger star, about 8 to 40 times the mass of the Sun. Aristotle's protostar will form a smaller star about the mass of the Sun, a mid-range mass for a star.

Act 2 Scene 2

This next scene is where the magic happens. This is where the star 'comes of age'. It develops from the adolescence of a protostar, into a fully fledged adult star.

Once the protostar reaches temperatures of over 10 million Kelvin, the first of many reactions in nuclear fusion occurs. The entire process of nuclear fusion will be discussed in the next chapter, for now we just need to know that it occurs. This is when the star is said to be 'self sufficient'.

Now that nuclear fusion is in play, the star begins to emit heat and light energy. Once again, an equilibrium is reached between the inward force of gravity causing the nuclear fusion, and the heat and pressure from said fusion pushing outwards, preventing the star from collapsing in on itself.

So now we have two new stars in our story. Aristotle forms an average star, and Plato forms a massive star. This is when our story regarding these two different regions starts to diverge. But for now, nuclear fusion will keep the stars burning for at least another 30 million years.

However, this is just what occurs if a protostar has sufficient mass and temperature for nuclear fusion to kick in. What happens if the mass is not high enough for nuclear fusion to occur? What becomes of the protostar?

Act 2 Scene 3

Let us move away from Aristotle and Plato for a scene. We will consider a much smaller molecular cloud. This molecular cloud collapses and forms a very low mass protostar. If the protostar is less than 0.08 times the mass of the Sun (solar masses is the term for this comparison), it will not form a star in which nuclear fusion occurs.

Instead, the protostar forms a brown dwarf. To put the mass of brown dwarfs, and therefore stars, into perspective, I will compare them to the largest planets. The largest planet in our solar system, Jupiter, is roughly 15 to 80 times lighter than a brown dwarf. And a brown dwarf is the smallest possible star that we can get.

Brown dwarfs do not undergo nuclear fusion, but some of them can still fuse together atoms of deuterium (an isotope of hydrogen) to emit some light and heat. Oddly enough, some brown dwarfs could have exoplanets orbiting them. However, the possibility of life and living on these planets is very thin due to a number of factors.

Brown dwarfs will eventually cool down and become dark balls of cold gas. This process can take billions of years, and the resulting star is called a black dwarf. Since there is no nuclear fusion, it does not have a traditional star death, but like everything else in the universe, it will lose all its heat through radiation and ultimately become a cold and barren wasteland. Anyways, enough about brown dwarfs, let us revisit our friends Aristotle and Plato, and see how their stars are faring 30 million years later.

Act 3 Scene 1

It is now 14th May in our Cosmic Calendar. A whole day has passed since the collapse of the molecular clouds, formation of protostars, and start of nuclear fusion. Our two stars have been burning brightly for 30 million years. However, for one of them, time is nearly up.

Since Plato's star is much more massive than Aristotle, it will run out of hydrogen faster and die earlier. In this case, we estimate it to be 30 million years. Aristotle's star however, will last for around 10 billion years, similar to our Sun. From this point on, we will talk about these two stars with a 9.97 billion year time gap in between. By the time Aristotle's star has run out of hydrogen, Plato's grandchildren and great grandchildren are just now entering the story. For the sake of showing what happens to each star, I will talk about them simultaneously, but with the added time gap.

We turn to Plato's star first. Once nuclear fusion has ceased, and all the hydrogen in the core has been used up, it evolves into a red supergiant. The two forces that have been in equilibrium for millions of years, gravity pushing in and pressure pushing out, are finally unbalanced. Moving towards the core (which now consists of helium), the gravitational force increases and the temperature rises. Moving away from the core, the outwards pressure dominates, due to the remaining hydrogen outside the core being fused, causing the outer layers to expand and increasing the size of the star.

In the core of Plato's star, a new form of nuclear fusion now occurs. Because gravity, density, and temperature is much higher than it has ever been, helium atoms will now fuse to form carbon and oxygen. This fusion will carry on in the core, forming heavier elements until it reaches iron. Iron is the last element formed in the core of a red supergiant.

Moving into the future, we come across Aristotle's star in its final stages of nuclear fusion. Unlike Plato's star, this star is much smaller and thus only forms a red giant, when hydrogen in the core is depleted. This red giant is formed in almost similar circumstances as the red supergiant. The star exhausts its core hydrogen and consists of a dense helium core, and the hydrogen outside the core carries on fusion causing the outer layers to expand and the star to increase in size.

This is a phenomenon that will happen to the Sun. In around 5 billion years, the Sun will become a red giant and swell up to at least 500 times its current size. This will completely engulf Mercury and Venus, and raise the temperatures on Earth a hundred fold. The oceans will be completely boiled off, and all life on Earth will cease to exist. Fortunately, we have around 4 to 5 billion years to come up with a solution to this dilemma, and potentially colonise another planet or star system.

Once Aristotle's star reaches its red giant phase, helium fusion begins in the core similar to the red supergiant of Plato's star. The key difference between these two is that in the red giant, fusion stops at carbon and oxygen. Temperature and density is not high enough to initiate fusion of heavier elements, and carbon/oxygen are the last elements formed in these stars. Anything between 0.3 and 8 solar masses forms a red giant, anything between 8 and 40 solar masses forms a red supergiant. Our two stars now enter very different phases of their lives.

Act 3 Scene 2

Plato's star will only stay in the red supergiant phase for around 1 to 2 million years. The end of this phase occurs when the core is entirely made up of iron. No more nuclear fusion will occur at this point. However, iron is only the 56th element in the periodic table. How will the other elements heavier than iron form? This is where the largest explosion in the universe enters our story.

Stars, Planets, and Everything in Between

Since there is no more energy being produced in the core by fusion, there is no outwards force to balance the inwards pull of gravity. At this stage, the core is only supported by electron degeneracy pressure. Essentially, the negative charge of the electrons causes them to repel each other and prevent them from moving any closer. However, once the core's mass exceeds 1.4 solar masses (the Chandrasekhar limit) degeneracy pressure can no longer support it and all hell breaks loose.

Gravity finally wins its millions of years long battle, and the outer layers of the core collapse inwards producing high energy gamma-rays (a form of electromagnetic radiation). This radiation will later ionise the surrounding cloud of matter that is produced. Inside the core, conditions are so extreme that protons and electrons will merge to form neutrons. The collapse of the core is halted at this stage by the neutron degeneracy pressure. When the collapse stops, the infalling matter rebounds (imagine throwing a ball at a brick wall and having it come back at you), producing a shock wave that propagates outwards. The shock wave blasts the material away, leaving only a degenerate core.

The rebounding matter is bombarded by extremely energetic neutrons, which they capture and use to produce heavier elements including all the radioactive elements. In fact, the reason these elements are radioactive is due to these neutron-proton reactions. We will explore the chemistry of these elements in the next chapter. We see that most of the elements heavier than iron are actually manufactured in the shockwave from a supernova explosion.

So Plato's star goes out in a violent fireball, permeating the fabric of spacetime with all the elements that it has produced through its own fusion, and all the elements it produced through its own demise. But what happens to Aristotle's star? In comparison, Aristotle's star goes out with a bit of a whimper.

After a billion years as a red giant, Aristotle's star enters the penultimate stage of its life. Once carbon-oxygen fusion has ceased in the core of the red giant, the surrounding gases are blown outwards at high speeds. The surrounding cloud of gases will become ionised by ultraviolet radiation. This stage is called the planetary nebula.

In comparison to the supernova of Plato's star, the planetary nebula is a much gentler and slower shedding of material. It occurs over a few thousand years, compared to the supernova explosion which occurs in a split second, ejecting material at velocities close to light speed. Supernovae will also produce much more intense forms of radiation, ionising the surrounding cloud of elements to higher energies. Speaking of this cloud, the material produced by Plato's star consists of almost every possible element in our universe due to the majority of them being produced in the explosion itself. The planetary nebula however, does not have sufficient energy to produce heavier elements, and thus the material ejected consists of mainly hydrogen, helium, and a handful of heavier elements (oxygen, carbon). This is what happens to the surrounding material and most of the mass of the original star, but what happens to the remnant core? We enter the final stage of a star's life.

Act 3 Scene 3

Ultimately, Plato's star's final form depends on its mass. If its mass is between 1.4 to 2.2 solar masses, it will remain as a solid core of neutrons, called a neutron star. If its mass is between 2.2 to 2.9 solar masses, the neutron degeneracy pressure is insufficient to support it, and it collapses into a black hole.

Let us talk about neutron stars first. The star (which is basically just the core of the original star) is only around 10 km in radius, with a mass exceeding the mass of the Sun. This means that a neutron star is one of the densest objects in our universe. Imagine the mass of 500,000 Earths. Now take that 500,000 and imagine it being compressed to fit in the space separating Central Park and JFK International Airport. That is how dense a neutron star is. Needless to say, its gravitational force is one of the strongest in the universe. In fact, dropping a tiny rock from hand height will produce enough kinetic energy, due to gravitational acceleration, to produce an explosion similar to that of an atomic bomb.

Since the neutron star does not undergo fusion, it will gradually cool down, losing heat to the surrounding space. The cooler the star gets, the harder it is for us to detect, as the electromagnetic radiation emitted is significantly faint. Altogether, there are estimated to be around 1 billion neutron stars in the Milky Way galaxy alone.

If the mass of Plato's star was above 2.2 solar masses, it would turn into a black hole. This term is often thrown around in sci-fi books and movies as a form of wonder and spectacle, but what does it truly entail?

A black hole is essentially a tear in the fabric of spacetime. Think of spacetime as a piece of cloth, unlimited in size and forever expanding. A black hole's density, mass, and gravitational force is so extreme that it literally rips a section of the cloth, resulting in a singularity in the centre. The singularity is the central region of a black hole where all the matter that was previously in the core of the star is now concentrated. The size of the singularity is zero. It has no size as it is a literal tear in space. These conditions are akin to the Big Bang singularity, and is one of the reasons why scientists are trying so hard to unlock the secrets of quantum gravity. A theory of quantum gravity will explain the final structure and reality of black holes, and potentially the reason for our entire universe existing.

Among the many theories of the singularity, my favourite is the theory of white holes. This theory states that the white hole is the converse of a black hole. The way a black hole prevents all matter from escaping, the white hole allows all matter to escape. It is basically a portal between the fabric of spacetime. The black hole sucks everything in, and the white hole spits it all out. Maybe our universe was created by such a white hole? It certainly makes for a very interesting conversation.

The other part of a black hole is the event horizon. The event horizon is a 'point of no return'. Anything that enters the event horizon, even light, will fall into the singularity and will forever be trapped in the gravity of the black hole. Nothing ever escapes the event horizon.

Exploring the Essence of Everything

Some rotating black holes will have accretion disks. In fact, these accretion disks are the only way of us actually 'seeing' a black hole, in terms of visual light. In reality, we observe the superhot gas of the disk swirling around a seemingly empty region of space. This empty space is the black hole itself.

The following paragraph contains spoilers from the movie *Interstellar*. Proceed with caution.

Perhaps the most famous black hole in science fiction is Gargantua, from Christopher Nolan's *Interstellar*. In *Interstellar*, the sheer gravity of Gargantua causes significant time dilation on the astronauts near it, which causes them to experience one hour as seven years back on Earth. We will explore the consequences of time dilation in space travel in chapter 9.

Additionally, Cooper, the main character of the movie played by Matthew McConaughey, enters Gargantua during the climax of the movie. He experiences incredible visuals and eventually falls into a 4 dimensional tesseract built by future humans who have learnt how to manipulate time and space to their will. He uses this tesseract to send binary and morse messages, using gravity, to his daughter back on Earth in order to save humanity from a blight. Of course all of these events are purely hypothetical, but if you would like to know more details about this, I strongly recommend reading Kip Thornes's *The Science of Interstellar*.

When we last left Aristotle's star, it had just become a planetary nebula, shedding its material in a colourful cloud of elements. The central core star forms something much tamer than a neutron star or a black hole. The final stage of its life is as a white dwarf.

The main difference between the neutron star and white dwarf, is that the white dwarf does not have sufficient gravity to cause its electrons and protons to merge to form neutrons. It exists as a solid ball of carbon and oxygen, and is supported by the electron degeneracy pressure.

Our Sun will eventually form into a white dwarf at the end of its life. Look out at the Sun right now. A massive ball of burning hydrogen. Emitting heat and light conducive to life on Earth. In 7 billion years, all that will exist of the Sun will be a sphere the size of the moon, cooling down over time.

Theoretically, the white dwarf will actually become a black dwarf once all its heat has been lost. However, the time frame for the formation of a black dwarf is so long, that it is actually longer than the age of the universe. The minimum amount of time for the Sun to become a black dwarf is one quadrillion years. Maybe I will live to see it one day.

And so, we reach the end of our journey. And what a journey it was. The collapse of the molecular clouds, the formation of the protostars, the first fusion reactions, the swelling up of the stars, the cosmic explosion of a supernova or the meek expansion of the planetary nebula, and finally the extremities of a neutron star and black hole in comparison to the tameness of a white dwarf.

However, before we can end this subsection we must allow things to come full circle. This is the end of the journey for these stars, but it is just the beginning for the next generation of stars.

Act 4 Scene 1

Plato's star and Aristotle's star left three gifts for the next generation. The first was the supernova remnant and the planetary nebula. You probably remember the significance of these clouds from the section on nebulae.

The second gift was the electromagnetic radiation from the ejection of material by the star. This electromagnetic radiation caused the surrounding material to become ionised. As previously mentioned, the supernova produces the more energetic gamma rays and the planetary nebula produces the less energetic ultraviolet radiation. Be that as it may, this ionisation of the surrounding material is crucial in development for the next generation of stars.

Granted, without the ionisation, the clouds of material would accrete matter over time and eventually collapse to restart the cycle all over again anyways. The ionised nebulae will just make it much easier to accrete material and collapse, speeding the entire process a hundred fold. Without this ionisation, the stars would take much longer to form.

The third gift was the surrounding material itself. Plato's star left a far more diverse set of elements which will cause the planets and moons formed from its supernova remnant to contain heavier elements potentially conducive to life. Aristotle's star leaves behind fewer elements, but over time and a few more generations, these elements will eventually increase in variety, enriching the entire universe with a whole spectrum of different elements.

Thus, the great cycle of life goes on and on. It has been going on for billions of years before our existence, and it will continue going on for billions if not trillions of years after our existence. We give thanks to this cycle for producing everything in the universe, including us. As clichè and overused as it is, we are literally made from starstuff.

The Lightbulbs of Our Universe

In this next subchapter, I will focus on the different types of main sequence stars that exist in our universe. The main sequence stars are the stars that undergo nuclear fusion of hydrogen in their cores. Hence, I will not discuss the latter stages of stars such as red giants, red supergiants, white dwarfs, neutron stars, or black holes, since they have already been covered in the previous subchapter. This subchapter emphasises on the seven different types of main sequence stars in terms of their size, mass, luminosity, temperature, colour, lifespan, and abundance.

The main sequence stars are characterised by letters, numbers, and roman numerals. The letters provide the more distinct method of difference, the numbers fine tune these differences, and the roman numerals provide differences in luminosity. The complete table of star types and their change in characteristics is shown in the next page.

The seven main sequence star types are O, B, A, F, G, K, and M. Each letter is then divided into ten more types using numbers from 0 to 9.

Stars, Planets, and Everything in Between

Finally, each letter is divided into seven roman numerals from I to VII. Fear not, we will not talk about all 490 different star types, we will just show trends and how the characteristics change across the spectrum of stars.

A famous mnemonic used to remember the seven types of main sequence stars is *"Oh Be A Fine Guy/Girl, Kiss Me!"*. Not a very subtle bunch these scientists.

Type	Size (solar radius)	Mass (solar mass)	Luminosity (solar luminosity)	Temperature (Kelvin)	Colour	Lifespan (Billion of years)	Abundance
O	> 6.6	> 16	> 30,000	> 30,000	Blue	0.01 - 0.03	0.000003%
B	1.8 - 6.6	2.1 - 16	25 - 30,000	10,000 - 30,000	Deep Bluish White	0.03 - 0.3	0.12%
A	1.4 - 1.8	1.4 - 2.1	5 - 25	7,500 - 10,000	Bluish White	0.3 - 2.0	0.61%
F	1.15 - 1.4	1.04 - 1.4	1.5 - 5	6,000 - 7,500	White	2.0 - 6.0	3.0%
G	0.96 - 1.15	0.8 - 1.04	0.6 - 1.5	5,200 - 6,000	Yellowish White	6.0 - 14.0	7.6%
K	0.7 - 0.96	0.45 - 0.8	0.08 - 0.6	3,700 - 5,200	Yellowish Orange	14.0 - 75.0	12.0%
M	< 0.7	0.08 - 0.45	< 0.08	2,400 - 3,700	Red	> 75.0	76%

Exploring the Essence of Everything

Firstly, we will see how the sizes of the stars change from O all the way to M. The unit of comparison will be in solar radius, how big the radius of the star is in comparison to our Sun. The size of the stars decreases from O to M. The minimum threshold for O type stars is a solar radius of more than 6.6 solar radii, and the maximum threshold for M type stars is less than 0.7 solar radii.

Next, we move on to the mass of the stars. We will use the familiar unit of solar masses to reference the mass of the stars. Similar to size, the mass of the stars decreases from O to M. The O type stars are usually more than 16 solar masses, and the M type stars are anything between 0.45 to 0.08 solar masses. Remember, anything less than 0.08 solar masses and the star is not able to carry out fusion and lives its life as a brown dwarf.

A pattern is starting to emerge here. Clearly there is some correlation between the size and mass of a star. Let us investigate further. The luminosity of the stars also decreases from O to M. The O type stars have luminosities northwards of 30,000 solar luminosity, while the M type stars have luminosities southwards of 0.08 solar luminosity.

Alright. Size, mass, and now luminosity. It only makes sense that temperature should be next. The two main features of a star that immediately springs to mind is the light and heat energy it emits. So it is no surprise to find out that even temperature decreases from O to M. O type stars radiate temperatures of more than 30,000 Kelvin, and M type stars radiate temperatures of between 2,400 - 3,700 Kelvin. In fact, the temperatures of stars differ so much that the aforementioned numbers between 0 and 9 (0 being the hottest and 9 being the coolest) are used to further whittle down these temperatures into finer distinction.

Another parameter used to define the temperature of a star is actually an everyday phenomenon. The spectrum of colours that we observe ranges from the high frequency, high energy, blue light, to the low frequency, low energy, red light. The same can be said of stars. Stars with higher temperatures will naturally produce electromagnetic radiation with higher energies. As a result, the hotter the star, the bluer it is. The spectrum changes from blue to white to yellow and finally to red as we move down the types from O to M.

The final piece of the jigsaw. The lifespan of a main sequence star. This is where all the previous classifications come together to form one single relationship to define the stars. Our trend finally reverses itself at the death. Before this, all the parameters of a main sequence star decreased from O to M. The lifespan of a star however, increases from O to M. The shortest lifespan of O type stars range from 3 to 10 million years, while the largest lifespan of M type stars range from 1 to 10 trillion years. This means that not a single M type star has ever died in our universe.

So how does the size, mass, luminosity, and temperature of a star correlate with its lifespan? Well, let's take it step by step. As the size of the star increases, more hydrogen atoms can fit inside it. The more hydrogen atoms inside a star, the heavier its mass. All ok so far. Of course, if one talks about mass one also has to talk about gravity. These massive stars will have strong gravitational forces which will cause temperatures in the core to increase. This increase in temperature will then speed up the process of nuclear fusion in the core. Hydrogen particles are colliding much more frequently and violently, and are also used up more quickly. The higher the rate of fusion, the more light and heat energy is emitted by the star. We see this as the heavier and larger stars also have higher luminosities and temperatures. Finally, as hydrogen is used up at a faster rate, the star depletes its hydrogen reserve quickly and dies in a few million years.

This is all in comparison with the rest of the star types. As size and mass decreases from O to M, the amount of hydrogen and strength of gravity in the core decreases. Thus, hydrogen will be used up slower, causing less heat to be radiated and the star to have a dimmer luminosity. Ultimately, due to the slow usage of hydrogen, the star is able to survive for a few trillion years.

The final characteristic is the abundance of each type of star. The O type star has an abundance of just 0.00003% of all main sequence stars, while the M type star has an abundance of 76% of all main sequence stars. The majority of stars are small, dim, red, and long lasting.

Exploring the Essence of Everything

These characteristics will be crucial when we discuss the suitability of a planet in supporting life. We will have to take into account the lifespan of the star, its temperature, and its luminosity. Will there be sufficient time for evolution to change tiny life forms into intelligent life forms? Will temperatures be conducive to liquid water? Will the brightness and radiation affect the mutation of potential life on the planet?

Our own Sun has a spectral class of G2V. This means that the Sun is a G type star, or a yellow dwarf. It has a temperature of around 5,600 Kelvin. The 2 means that it is on the hotter end of the spectrum for a G type star. The V in roman numerals just shows that its luminosity is that of a main sequence star. All stars are characterised using this same notation. A letter for star type, a number for temperature in that star type, and a roman numeral for luminosity type.

Thus, we reach the end of our lengthy discussion on stars. We have learnt how these beasts of heat, energy, and light are born, how fusion is initiated, how they survive for millions and billions of years, the different types of hydrogen burning stars, and the various ways in which a star dies. The final two subchapters of this chapter will focus on the much smaller element of a stellar system, but arguably the most familiar and important to us. We will take a closer look at the formation, life, movement, and characteristics of the celestial planets.

Your Home Is Here

Like everything else in the universe, the formation of a planet starts with miniscule differences in density and gravity. The entire structure and basis of our planet originated from a single 0.1 mm grain of dust. To put this length into perspective, it is similar to the width of a strand of hair, the height of a sheet of paper, and is the absolute limit of human visibility with the naked eye. Imagine the largest structures in our world. The peaks of Everest. The skyscrapers of New York City. The statues and relics of ancient civilisations. Everything obscenely large originated from something obscenely small.

Stars, Planets, and Everything in Between

In fact, everything that you see around you, originated from an interstellar molecular cloud. We have already covered the collapse and formation of the central star, but what of the remainder of material in the cloud? Obviously not all the material goes into the making of the star.

Once the central protostar is formed, the rest of the material swirls around it in the form of a protoplanetary disk. This disk, also called an accretion disk, is flattened due to the same laws of conservation of angular momentum that caused the flattening of our Milky Way and the Earth's poles. This is why all the planets in the solar system lie on the same plane.

The protoplanetary disk has a few distinct regions which are key to the composition of the planets. The inner region closer to the protostar contains heavier metals and elements, with many rocks and minerals present. This is where our rocky, terrestrial planets are formed. Some distance past this region is the frost line. The frost line is the minimum distance from the protostar where the temperature is low enough for gases to condense into solid compounds. This is why the gas giant planets (Jupiter, Saturn, Uranus, Neptune) are solid even though they are made up of gas and ice.

So we start with a single grain of dust. It accretes another dust grain. And another. And another. And another. Long story short, you know how this process goes. Eventually, these grains clump into larger bodies called planetesimals of around 10 km in size. However, don't think that this process is instantaneous and that there are planets popping up throughout the universe like so many pieces of popcorn. This is an extremely drawn out and boring process. This is a growth of a few centimetres every year. The entire process takes a few million years. In the time it takes for an O type star to be born, start fusion, exhaust its supply of hydrogen, turn into a supergiant, go supernova, and finally come to rest as a neutron star or black hole, our miniscule planetesimal is not even close to being fully formed.

Further growth is possible because these mini planets collide with each other frequently, and with large velocities. It is these collisions which formed many of the moons orbiting the planets today. After a few million years, the central star's solar wind would have cleared all the dust and gas in the protoplanetary disk, blowing it into interstellar space and halting the development of the planets. But why are the rocky planets much smaller than the icy gas giants?

The main reason for the differences in size lie in the composition of the planets. As previously mentioned, the rocky planets are composed of much heavier elements and metal. We know that the original molecular cloud only contained 1% of elements heavier than hydrogen and helium by particle count. Most of the hydrogen and helium in the protoplanetary disk would have been blown further out by the radiation and heat from the central environment. Hence, the main elements that could have made up these rocky planets were the heavier, less abundant elements. As a result, the terrestrial planets are much smaller in size. Even though hydrogen is by far the most abundant element in the universe, it is actually quite sparse on Earth compared to the other elements.

The converse of this occurs with the much bigger gas giants. Since most of the hydrogen and helium has been blown out past the frost line, these gases along with a few other gaseous compounds, will condense into solids. These solids will then go through the similar process of accretion to form planets. Since the abundance of these light elements and gases is much higher than the elements closer in, there is simply more material to accrete and the planets formed are much bigger. In fact, around 99% of the mass orbiting the Sun is found in the four gas giant planets. However, as a consequence to this, the outer planets are much less dense and the inner planets are much more dense.

We expect these trends of planet formation in our own solar system to be extrapolated throughout the universe. This pattern should be recognizable everywhere else, and would make for easier detection of exoplanets. However, there are certain cases where planets with the size and composition of Jupiter orbit their stars at closer distances than Mercury orbits our own Sun. We will entertain this idea in the last subchapter.

Finally, we have to talk about the lack of a certain planet in the room. Why was Pluto shunned from our planetary encyclopaedia and cast out to be another one of the random spheroids orbiting our Sun? What can we classify as a 'planet', and what doesn't quite meet the criteria?

Stars, Planets, and Everything in Between

The International Astronomical Union defines three specific rules that a celestial body must meet in order to be deemed a planet.

1. It must orbit a star.

2. It must be big enough to have sufficient gravity to force it into a spherical shape.

3. It must be big enough that its gravity has cleared away any other objects of a similar size near its orbit around the star.

The issues faced by Pluto first started in 2005 when a tenth planet was found. This planet, named Eris, was found in the vicinity of Neptune and appeared to be slightly larger than Pluto. This sparked a debate amongst astronomers as to what can actually be deemed a planet.

Pluto passes the first two rules with distinction. It orbits the Sun, and is quite clearly spherical in shape. However, it is the third rule where Pluto, and many other celestial objects, fall short. Most of them have other objects of similar size in their vicinity. Pluto still has a lot of asteroids and moons that are close to its size in its surrounding environment. This same rule applied to Eris, and a new class of planets was born.

These so-called 'dwarf planets' meet the first two criteria for a planet but do not have to meet the third. In addition, the dwarf planets cannot be natural satellites, in order to differentiate them from the moons of the other planets. Once again, these rules for planets and dwarf planets are applicable throughout the universe.

Now that the planets have formed and have a fixed composition, we move into our final subchapter. We will discuss the orbits of these planets, the phenomenon of planetary migration, and the workings of one Johannes Kepler.

Exploring the Essence of Everything

The Great Merry-Go-Round

The next stop on our cosmic voyage of the stars and planets takes us to the town of Weil der Stadt, in Stuttgart, Germany, during the end of the year 1571. Johannes Kepler was born into a fairly large family with three siblings, and spent the majority of his childhood years living in his grandfather's inn. He often impressed travellers at the inn with his phenomenal mathematical ability, and would go on to become the greatest mathematical and scientific mind of the century.

Young Kepler took a keen interest in the wonders of the heavens and its celestial elements, observing comets, lunar eclipses, and a handful of the stars. A bout of smallpox left him with weak vision and crippled hands, stunting his career in the observational aspects of astronomy.

At the time of his adolescence, the debate of the geocentric model vs. the heliocentric model was all the rage. Kepler studied both and naturally took a liking to the heliocentric model of Copernicus, stating that the Sun, not the Earth, was the centre of the universe. This inspired many of his studies on celestial bodies. Similar to Aristotle, the works and labours of Kepler could be penned into an entire book of its own. For our purposes, we will focus on the development of his three laws of planetary motion.

Kepler altered the laws of planetary movement formulated by Copernicus, fine-tuning the orbits from circular motions and constant velocities to elliptical trajectories with varying velocities. This left him with three primary rules for the movement of planets around a star (the original laws were based on the movement of Mars around the Sun, but the laws can be extrapolated to include all planets and all stars in the universe).

1. The orbit of a planet is an ellipse with the star at one of the two foci.

2. A line segment joining the planet and its star covers equal areas in equal time intervals.

3. The square of a planet's orbital period is proportional to the cube of the length of its semi-major axis of its orbit.

Clearly a lot is being said here. Let us break it down in layman's terms. The first rule states that the motion of a planet around its star is not a perfect circle. Rather, it follows the shape of an ellipse. Think of an ellipse as an oval where the proportions and dimensions are equal if cut in half. It is essentially a symmetrical oval.

We have many types, or rather degrees, of ellipses. Some are closer to being perfect circles, some are so elongated that the orbit of the planet takes it billions of miles from its star at the farthest point. The degree of an ellipse is measured by its eccentricity. The eccentricity of an ellipse, and therefore a planet's orbit, is between 0 and 1. An eccentricity closer to 0 means that the orbit is almost perfectly circular, and an eccentricity closer to 1 means that the orbit is closer to that of a parabola. A planet with a parabolic orbit will not complete a full orbit around the star and will be ejected out of the stellar system.

Another key point about elliptical orbits is the distance of the star from the planet. In a perfect circle, the centre is always equidistant from the edge. As the orbit gets less circular, this will alter the position of the star relative to the planet. The star is no longer perfectly at the centre. For objects with high eccentricities, the star is located at one end of its orbit, meaning that the distance of the planet from the star fluctuates wildly, as it is away from the star for the majority of its orbit.

For reference, the planets in our solar system have eccentricities from 0.006 (Venus) to 0.093 (Mars). The Earth has an eccentricity of 0.016. This means that the planets in our solar system have very nearly perfect circular orbits. This also means that the planets are roughly equidistant with the Sun at all times, and that there isn't much difference between their perihelion and aphelion. This is one of the many curiosities of our solar system. Why should we be blessed with these fortuitous circular orbits? If the Earth had a wildly elliptical orbit, its distance from the Sun would fluctuate dramatically causing the harshest summers and the coldest winters. Instead, our temperatures are relatively stable due to this low eccentricity. Is it a condition for life to evolve on planets where the orbit is almost circular and the temperature is stable?

Not all objects in our solar system have low eccentricities. Halley's Comet, first discovered by English astronomer Edmond Halley in 1758, has an eccentricity of 0.967. This means that it follows a highly elliptical orbit, with an orbital period of roughly 77 years. Halley's comet will be visible again sometime in the year 2061. Mark your calendars. The comet has such an elliptical orbit, that at its closest distance from the Sun (perihelion) it is only 88 million km away, but at its farthest distance (aphelion) it is 5.2 billion km away, roughly the same distance away as Pluto. Imagine if Pluto was as close to the Sun as Venus is every 77 years!

Kepler's second law follows on from the first law. It states that the closer a planet is to a star, the faster it will travel. This also ties in with the laws of gravity. As the planet gets closer to the star, gravitational force increases. This gravitational potential is converted into the kinetic force of the planet, making it move faster. Objects that are attracted by a higher force of gravity will move faster through space, that is one of the many consequences from Einstein's Theory of Relativity.

For planets with almost circular orbits, such as the planets of our solar system, this change in speed is almost non-evident, since the gravitational force between the planet and the Sun is always constant. Another benefit of Kepler's laws to be thankful for. Could you imagine if we were on a planet with a higher eccentricity, forever accelerating as it got closer to its star and decelerating as it moved away? The force of inertia experienced would be catastrophic and almost all structures and life would be wiped out with this continuous braking and speeding up.

The third and final law of planetary motion states that, the farther a planet is from its star, the slower its orbital speed and vice versa. This is probably the easiest of the three laws to understand. It simply means that objects further out will take a longer time to complete a single orbit of its star. This is once again tied to the gravitational effect that a star has on its planets, mentioned in the second law. Since the planet is further out, it receives less gravitational potential from the star, and subsequently less kinetic force making it travel slower and further around the star.

The phenomenon that once baffled scientists when it came to a planet's orbit is the curious case of planetary migration. Planetary migration was first brought to light when a number of Jupiter-sized exoplanets were detected, and it was found that they were extremely close to their parent star, with orbital periods of just a few days. Imagine an entire year occurring in the matter of a few days. That's how close these planets were to their star. Of course, there wouldn't be enough matter for these huge planets to accrete right next to their star, and isn't Jupiter supposed to be made out of ice and gas anyways? This seemed to be the greatest teleportation trick in the universe. How does a planet the size of Jupiter move to within an inch of its star?

There isn't a 'one size fits all' answer to this question. In reality, a number of situations could cause planetary migration. To provide the details that affect us and our solar system, we will just discuss the main cause, but as always, additional reading material and sources are provided at the end of the book. Let us see how this magic movement of the planets has affected our own solar system.

The main cause of planetary migration is due to disk migration. Interactions between the planet and the protoplanetary disk causes density variations and a loss of angular momentum. As the planet loses angular momentum, it starts to gradually spiral in towards the star. For some systems, this spiralling occurs until the star has completely trapped the Jupiter-sized planet in its vicinity, keeping it tidally locked and at arm's length. But clearly something stopped the migration of Jupiter in our solar system. To what do we owe this fortuitous braking?

We have to thank Saturn for halting Jupiter in its tracks. Allow me to elaborate further. Jupiter was not the only planet spiralling inwards at this time, in fact all the gas giants were on a similar crash course for the Sun. However, at one point of their journey Jupiter and Saturn were caught in a 2:1 orbital resonance lock. For every 2 orbits that Jupiter made around the Sun, Saturn would make one. And this ratio is constant even to this day. This orbital resonance stopped Jupiter and Saturn from spiralling further towards the Sun, and kept them in stable orbits. Uranus and Neptune were also halted by orbital resonance, theirs being in the ratio of 51:26 (very similar to the 2:1 of Jupiter and Saturn).

In some stellar systems, orbital resonance does not take place, or is broken by turbulence in the disk, interactions with other planets, or tidal forces from the star. Thus, the planet continues to fall inwards, causing us to detect them as Hot Jupiters.

The halting and permanence of Jupiter in its current orbit and position in the solar system has significant influence. So much so, that it has attained the moniker of 'vacuum cleaner of the solar system'.

Firstly, if Jupiter had not been stopped by Saturn, it would have continued on in its demolition run towards the Sun. The rocky planets of Mars, Mercury, Venus, and Earth would have been swallowed up by this marauding titan, or cast out altogether from the solar system. Perhaps the existence of Hot Jupiters in a stellar system is a sign that there is not much room for life to evolve in that system.

Jupiter eats up and deflects the asteroids and comets that venture directly towards the Sun, acting as a shield for the inner rocky planets. This is evident by the asteroid belt orbiting between Mars and Jupiter. Without the presence of a massive structure like Juptier (and to some extent Saturn), we would be facing many, many more asteroid impacts on Earth, some even akin to the one that wiped out the dinosaurs. The more asteroid impacts and extinction events, the less likely it is that life would have evolved to form the conscious, intelligent, humans of today.

Finally, we have reached the conclusion of our odyssey of the origins, formation, and structure, of not just the universe, but the many structures within. From the picturesque panoramas of the nebulae, to the scorching starlight of the many lightbulbs, to the andante accretion of the planets, we have learnt and discovered how the celestial objects in our universe take shape. We have gone back billions of years with analysis of the Cosmic Microwave Background, and the Big Bang Theory. We have seen how tiny quantum fluctuations in the fabric of spacetime can have the most profound and immense impact, with the formation of galaxies. We have explored the many types of stars, the birth of said stars, and the death and demise of these great furnaces. And we have seen how our place in the solar system came to be, bringing everything full circle back to the present day.

Now that we have a better understanding of the astrophysical elements of our universe, we shall take a closer look at the astrochemical aspects. We know how stars and planets came to be, from the differences in density and mass, and the helping hand of gravity, but what exactly forms the bonds and composition of all these objects? What are the building blocks of our universe? How do we go from the tiniest of protons and electrons to fully scaled stars and planets? What even is an atom? How does the existence of these elements permeate the fabric of spacetime? What is its importance? To answer all these questions and more, we embark on the next leg of our cosmic journey, from the stars to the supernovae.

The Big Bang theory was first proposed by Georges Lemaître in 1927. The Belgian priest, physicist, and astronomer presented theories on the expansion of the universe. (Katholieke Universiteit, Leuven)

Edwin Hubble proved Lemaître's theories when he discovered that the universe was expanding, after analysis showed that distant galaxies were receding at a faster rate. (Johan Hagemeyer)

Robert Woodrow Wilson (left) and Arno Penzias (right) discovered the Cosmic Microwave Background, which further strengthened the Big Bang theory. (NASA)

The Holmdel Horn Antenna, located on Crawford Hill in New Jersey, was used to detect the Cosmic Microwave Background. (NASA)

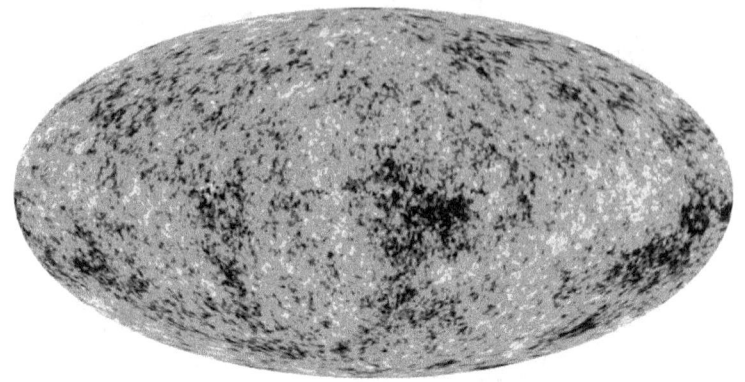

Temperature fluctuations in the Cosmic Microwave Background can be seen by the dark spots (hotter regions) and the light spots (colder regions). The hotter regions are less dense while the colder regions are more dense. These differences in density are caused by the primordial quantum fluctuations during inflation. (NASA/WMAP Science Team)

An image of the Milky Way taken above Chile, on 21st July 2007. Note the wide band of stars and dust clouds. The laser is pointing directly at the galactic centre. (ESO/Y.Beletsky)

An image of Sagittarius A*, the supermassive black hole in the heart of the Milky Way. The lines show signatures of a magnetic field around the shadow of the black hole. (Event Horizon Telescope collaboration)

The Orion Nebula is an emission nebula, a dense region of star formation. (NASA/ESA)

Messier 78 is a reflection nebula. The light from distant stars in the image is scattered by the dense dust clouds, making the entire nebula visible. (ESO/Igor Chekalin)

The Horsehead Nebula is a dark nebula. The dark regions in the nebula absorb and block light from background stars. (ESO)

The Trifid Nebula is rather unique as it shows properties of all three diffuse nebulae types. The central region is an emission nebula, the upper region is a reflection nebula, and the region towards the right is a dark nebula. (ESO)

The Ring Nebula is a typical planetary nebula, as a central star sheds its outer layers. (NASA/ESA/CSA/James Webb Space Telescope)

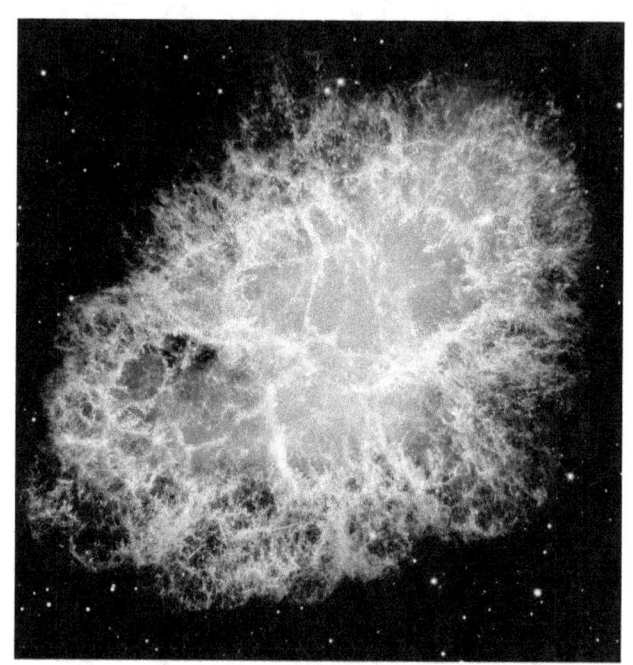

The Crab Nebula spans a distance of 6 light years, the nebula of a supernova remnant. This violent event was first recorded in 1054. (NASA/ESA)

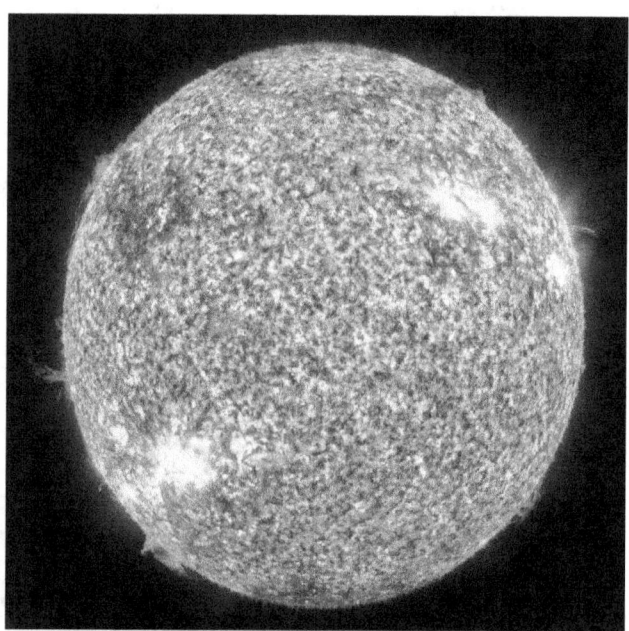

Our Sun is a G2V type star, with G2 showing that it is the second hottest version of a G-type star, and V representing a main sequence star. (NASA/SDO)

Johannes Kepler formulated the three laws of planetary motion, describing the orbit of a planet around its star. Kepler witnessed a supernova in 1604, and also refined several telescopes throughout his lifetime.
(August Köhler - Kepler Museum in Weil der Stadt)

2

Chemistry of the Universe

*"We are what happens when a primordial mixture of
hydrogen and helium evolves for so long,
that it begins to ask where it came from"*

Jill Tarter

Elementary, My Dear Watson

Let's start this chapter off with an exciting mental exercise, testing our powers of imagination, visualisation, and comprehension. This exercise will act as the basis and foundation of our understanding of chemistry for the rest of the chapter. It will also show us how every single substance of matter in our universe is the exact same at its core. Wherever you may be at this current moment, reading this sentence, take a look around and focus on a singular object. It doesn't have to be big, small, round, colourful, edible, living, etc. As long as you can see it, and are somewhat familiar with what it's made from, it will do.

As I am writing this, I have a particularly delectable grilled cheese sandwich next to me (I'm sure everyone is familiar with those late night cravings). I will use this sandwich as my example in the exercise but feel free to focus on whatever you desire. Now, I know that this sandwich is composed of a mixture of bread, cheese, butter, and whatever else I fancied to put in there. Let me focus on the bread then (I could easily focus on the cheese or butter, it makes absolutely no difference in the end, as you shall see).

The bread is composed of flour, yeast, water, and salt, maybe some sugar mixed in there as well. Alright let us move on to the flour then. As you can see, what we are essentially doing here is breaking things down into smaller and smaller sections, and seeing how far we can go. Do follow along with your item of choice. If you're unsure as to its composition, a simple google search will do. As stated earlier, the end result will be exactly the same.

Ok so my figurative microscope has gone further into my sandwich and detected flour. The ingredients on the loaf of bread state that this is wheat flour, composed of mainly carbohydrates, with trace amounts of protein, fat, vitamins, and minerals. (Looking back at it, finding the actual type of flour was completely pointless since all of them are made up of carbohydrates anyway). If you selected an edible or living organism as your reference object for this exercise, you will eventually

Chemistry of the Universe

reach this stage as well. A mixture of the basic biochemical molecules. This is pretty much akin to going down a rabbit hole of wikipedia pages, deeper and deeper into the constituents of my sandwich.

Carbohydrates next. Carbohydrates are a molecule with the elements carbon, hydrogen, and oxygen, in the ratio 1:2:1. If you have selected a non-living object, you will also reach this point at some stage. This is the point where all our objects share a common ground. In fact, the first important common ground. They are all composed of elements. Some may be elements on their own (if you had chosen a diamond as your object for example), some may be compounds, some may be mixtures.

Into the world of the elements, I continue my cross-section of my grilled cheese with carbon, the element of life (on Earth anyways!). Feel free to choose any element in your object, you can search up its proton/electron number if you want but it's really not necessary. Carbon is composed of 6 protons, 6 neutrons, and 6 electrons. How boring. In fact, any element that you have chosen will have some number of protons, neutrons, and electrons (the only exception being hydrogen-1, which is why I didn't choose it).

Most high school syllabus of chemistry and subatomic particles stop here, so this should still be familiar to most of you. We have all reached the same stage. It doesn't matter whether you choose your dog, a skyscraper, a bit of grass, your neighbours car, whatever. We have all reached the domain of protons, neutrons, and electrons.

So our final step in this exercise is to break down these subatomic particles. But the question remains, how far down can we go? How much smaller can I split my grilled cheese? I have gone from sandwich to bread to flour to carbs to carbon to protons. What's next? This is the story of our elementary particles, the basis of every piece of matter that has ever existed in our universe.

When we were last discussing the stages of the Big Bang, we stopped at 10^{-10} to 10^{-4} seconds. At this stage, our universe would have been permeated with a quark-gluon plasma. What does this mean? To explain quarks and gluons, we have to first break down the different types of elementary particles, and their subsections. Don't worry, we won't go into too much detail, but additional reading material will be provided at the end of the book should you wish to know more.

I realise that there will be many unfamiliar names and terms in relation to the elementary particles in this section. Use the two diagrams on page 89 to get a better understanding on how all these particles are connected.

At the present time, there are 17 known and observed elementary particles. These 17 particles can be split into two groups, fermions and bosons. Fermions are our 'matter' particles. Bosons are the particles which give certain characteristics to everyday matter. To know whether a particle is a fermion or boson, we look at its spin number. The concepts of 'spin number' exceed the syllabus of this book, but the key point to remember is that spin number can either be half integers ($\frac{1}{2}$, $\frac{3}{2}$, $\frac{5}{2}$) etc, or whole integers (0, 1, 2). If a particle has a half spin integer, it is a fermion, if it has a whole spin integer, it is a boson.

Let us explore fermions first. Fermions can be further divided into two groups. Leptons and quarks. The main difference between the two is that leptons interact with the weak nuclear force only, whereas quarks interact with both the strong and weak nuclear force. The technicalities of these forces will be discussed in the next subchapter. There are 3 types of leptons; electrons, muons and taus. Each of these three leptons are negatively charged but have neutrally charged counterparts called neutrinos; electron neutrino, muon neutrino, and tau neutrino respectively. Quarks on the other hand, have six types, or 'flavours'. Up, down, charm, strange, top, and bottom. These 12 fermions (6 leptons and 6 quarks) are the basis for all matter in the universe.

To complement the 12 fermions, we have 5 boson particles. These 5 bosons can be split into two groups; 4 gauge bosons, and 1 scalar boson. The gauge bosons are our 'force carriers'. They act as the particles transmitting and conveying the four fundamental forces of our universe. The photon carries the electromagnetic force, the gluon carries the strong nuclear force, the W and Z bosons carry the weak nuclear force, and the graviton carries the gravitational force.

Hold on, that makes 5 gauge bosons instead of 4. The reason we say 4 is because the existence of gravitons have thus far not been confirmed with direct observation and evidence. They remain a hypothetical particle. However, it remains to be said, if there are force carrier particles for the other three forces, then there should be one for gravity as well. The main reason that the graviton has not been detected yet is due to the gravitational force having the weakest strength compared to the other three forces.

The single scalar boson is called the Higgs boson. Only discovered in 2012, the Higgs boson is associated with the Higgs field, which gives mass to all the other particles. So we see how these 5 (or 6?) bosons provide the defining forces and quantities for the rest of the particles.

If fermions are the yin, then bosons are certainly the yang. Both complement each other and cannot exist without the other, or even if they did, they would not amount to anything. For there to be matter, and eventually life, these two types of elementary particles must merge in a grand marriage, spanning billions of light years in space and time.

How then do we go from elementary particles to our protons, neutrons, and electrons? Well we now know that electrons are a type of lepton. What about protons and neutrons? For that, we have to shine the spotlight on the quark and the gluon.

Quarks and gluons work together very well. This refers back to our 'quark-gluon plasma' mentioned a couple pages earlier. The quark-gluon plasma is described as a state of matter where the baryonic matter of hadrons are freed from their strong attraction between each other, and are free to permeate the universe individually. Alright, a couple new words here. First I will have to explain the mutualistic relationship between quarks and gluons.

As we know, quarks are the 'matter' particles, and gluons are the 'force carriers', specifically the strong nuclear force. This strong nuclear force enables the quarks to be attracted to one another. A hadron is any type of subatomic particle built from two or more quarks. There are two types of hadrons; baryons and mesons.

Baryons consist of three quarks, each with a half spin integer, and are thus classed as fermions. The two most famous baryons are our protons and neutrons. See how it all fits together now? Protons and neutrons are made up of three quarks, held together by the strong force, which is carried by gluons. The charge of the quarks inside the protons and neutron determine the charge of the subatomic particle itself. The two up quarks and one down quark in a proton results in a charge of +1 while the neutron consists of two down quarks and one up quark, resulting in a charge of 0.

Mesons consist of two quarks, each with a half spin integer. Since there are only two quarks in a meson, the half spin integers cancel each other out, resulting in a meson with a whole spin integer; 0, 1, or 2, meaning that mesons are a type of boson. It might seem confusing as to how mesons, which are made up of quarks (a type of fermion), can end up being bosons. But remember, the main difference between a fermion and a boson is just its spin integer. Examples of mesons are pions and kaons.

We can finally complete our dissection of my grilled cheese, now that we have all the pieces of the puzzle. Starting with the whole sandwich, we go into the bread, which is made up of flour. Flour consists of carbohydrates, a molecule that contains the element carbon. Carbon has 6 protons, 6 neutrons, and 6 electrons. Electrons are a type of elementary particle on their own (lepton) and can't be broken down further. Protons and neutrons are baryons made up of three quarks each. Quarks are an elementary particle and can't be broken down further. So no matter what you started with, you will also end up with a mixture of quarks and electrons. But if we are all made up of the exact same thing, how are we any different from nonliving matter? Whether living or nonliving, we are all just a mixture of quarks and electrons with a few bosons thrown in there. Where do we draw the line between living and nonliving? A tale for another day.

Chemistry of the Universe

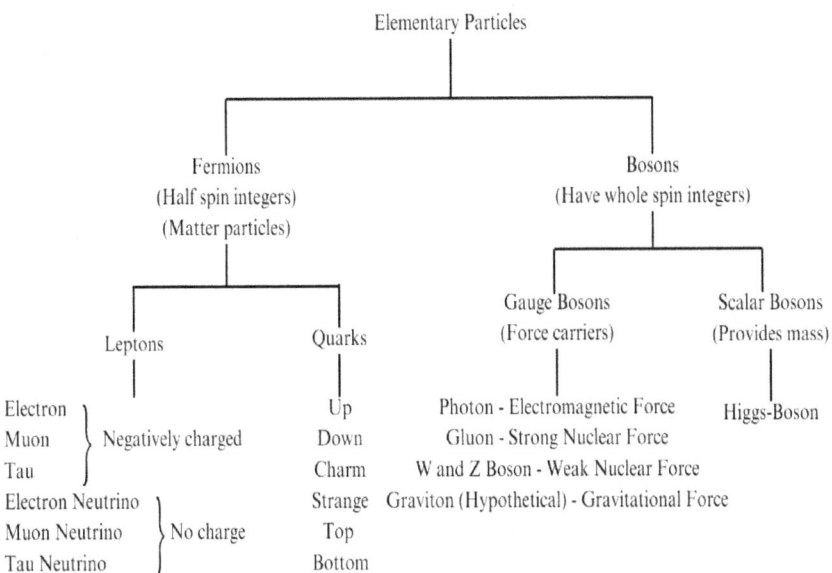

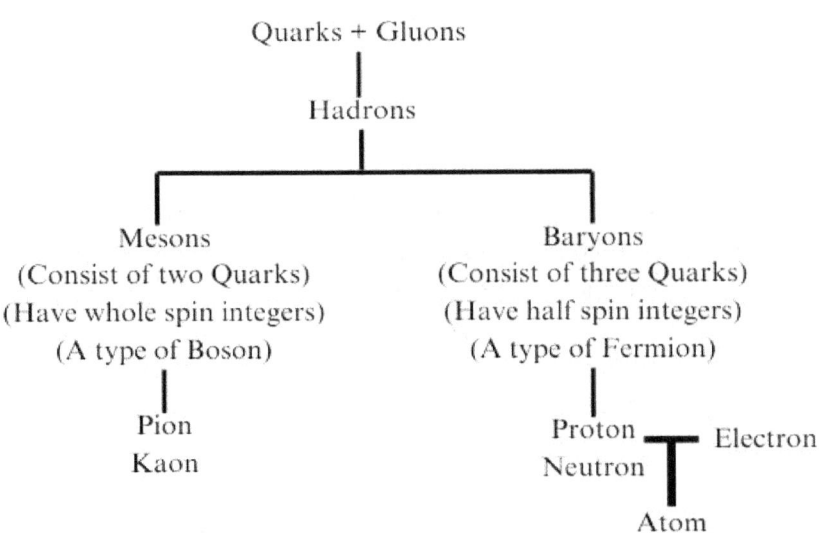

Exploring the Essence of Everything

The majority of these elementary particles are subject to extensive research and experimentation by scientists worldwide. Billions of dollars have been spent to build particle accelerators such as the Large Hadron Collider at CERN, the Stanford Linear Accelerator, and various synchrotrons around the world. A handful of mysteries and early universe conditions are trying to be solved and replicated, among them, recreating the primordial 'quark-gluon plasma', solving the asymmetry of matter and antimatter, finding extra dimensions predicted by models of string theory, and answering many questions regarding the four fundamental forces of our universe.

I have mentioned in passing about the existence of the four fundamental forces of nature. Let us now examine these mystical forces in more detail and see how they form the structure of our universe.

May the Forces Be With You

In a standard high school Physics class, you would have learnt that a force can be described as a pushing or pulling action. This definition is suitable for that level of education, and it is certainly not wrong, but if we extrapolate 'push' and 'pull' as well as all the other forces we are familiar with (friction, tension, applied, normal, etc), a simplified picture emerges. In this picture, every single force in our universe falls into one of four categories; the strong force, the weak force, the electromagnetic force, and the gravitational force. These four forces work together (and sometimes individually) to create all the possibilities in our universe. Our structure of matter, the sunlight that hits our skin, apples falling from trees, the ability to hold an object without it passing through your hand. Almost every single outcome in our universe is governed by these four fundamental forces.

There are many ways to separate these four forces. Initially, I will split them into two groups; the strong and weak nuclear force on one side, and the electromagnetic and gravitational force on the other. The strong and weak nuclear force produce forces at miniscule distances but with enormous magnitudes of strength, and are confined to work in the nucleus. These two forces dictate the nuclear properties of an atom. The electromagnetic and gravitational force produce long-range effects, with ranges spanning infinity, but decreasing in magnitude as the distance from the source increases.

We shall start with the strong nuclear force. This force was touched upon briefly in the previous section as being responsible for holding quarks together to form protons and neutrons. The strong force is carried by the gluon elementary particle, and its current theory is quantum chromodynamics (QCD). However, holding quarks together is not the only thing the strong force does.

We know from basic high school chemistry that the proton is positively charged and that the neutron is neutrally charged, or has no charge. We also know that like charges repel, and opposite charges attract. Finally, we know that in the nucleus of an atom, there only exists protons and neutrons. Electrons merely orbit the atom but are not actually present in the nucleus itself. With these three prerequisites, this begs the question; what is stopping the protons from repelling each other and flying out of the nucleus? If the nucleus is positively charged and like charges repel, what is responsible for keeping these protons together? This is where the strong force comes into play once again.

It is the strong force which is responsible for keeping protons bound in the nucleus of an atom, the same way that it keeps quarks bound in the protons and neutrons. Without the strong force, there would be no structure of matter whatsoever. No galaxies, no stars, no planets, no humans. You get the idea. Atoms themselves would not exist since protons would not be able to be held together, and the universe would become a giant sea of quarks (and the other elementary particles). The strong force is the ultimate super glue of our universe.

Additionally, the strong force also allows the protons and neutrons to coexist in the nucleus, without smashing into each other. How much attraction between protons is too much? At extremely small distances,

7^{-16}m, the strong force actually causes the protons and neutrons to repel each other.

In a shocking turn of events, it appears that the strong force is actually the strongest of the four fundamental forces. However, even the strong force has its limits.

Just because this force is responsible for holding nuclei and hence atoms together, doesn't mean that we can make these nuclei as big as we want, or that we can add as many protons as we desire. At a distance of 10^{-15}m, the strong force is 10^{38} times stronger than gravity. Although, if we move out to greater distances by making the nucleus bigger, the power of the strong force diminishes rapidly and is no longer strong enough to keep all the protons together. This is part of the reason why we can only have 118 elements, and why most of the larger ones are radioactive. As we add more protons and neutrons into the nucleus, the size increases. As the size increases, more force is needed to keep the protons together, and the force diminishes as the size gets bigger. This results in instability and the emission of alpha particles. We will cover discussions on radiation in more detail further on.

So finally, we have a better understanding of this mystical force. Initially, it acts as the glue to hold quarks together to form protons and neutrons. At extremely close distances, it prevents neutrons and protons from smashing into each other with repulsion. Move slightly further out and it binds the protons in the nucleus of an atom together, preventing them from flying out of the atom. And as the size of the atom increases, the force diminishes rapidly, placing a definite limit on the number of protons and neutrons in an atom.

As Newton said, for every action there is an opposite, but equal, reaction. If the strong force holds the subatomic particles in the nucleus together, there must be a force which allows the particles to escape the nucleus completely. This force is called the weak force.

In some respects, the weak force can be the most challenging of the four forces to explain simplistically. We can simplify the strong force as the force needed to hold positively charged protons together. We can simplify the electromagnetic force as the force between electrically charged particles. Gravity can be simplified as the mutual attraction between objects of mass. Saying that the weak force is responsible for radioactive decay is certainly not wrong, but it does require a bit of in depth explanation.

One form of radiation is called beta radiation. There are two types of beta radiation; beta minus, β^-, and beta plus, β^+.

β^- decay occurs when a neutron in the nucleus of an atom changes into a proton, an electron, and an electron antineutrino. Previously, we learnt that both neutrons and protons are examples of baryons; they are both made up of three quarks each. We also know that in a reaction, matter and energy cannot be created or destroyed. The total mass and energy before must equal the total mass and energy after. How then, does a neutron change into a proton, plus two other elementary particles? Has one of nature's most fundamental laws been violated?

To explain this phenomena, we must dig deeper into the constituents of a neutron and a proton. This is what I mean regarding the complexities of the weak force. Fear not, if you were following the previous section, most of this will be familiar.

There are six 'flavours' or types of quarks. Up, down, charm, strange, top, bottom. We know this. Each proton and neutron consists of three quarks each. This we know also. Protons and neutrons are made up of different flavours of quarks. Protons are made up of two up quarks, and one down quark, whereas neutrons are made up of two down quarks and one up quark. Again, this is known.

The next step arises when we talk about how the flavour of the quarks change. Remember, the only difference between the proton and neutron is that one has 2 up 1 down (proton), and the other has 1 up and 2 down (neutron). This causes the neutron to be slightly heavier than the protons. In beta minus decay, one down quark in the neutron is changed into an up quark. When this occurs, the neutron is converted into a proton. But like I said, where is the conservation of mass and energy?

When the down quark is converted into an up quark, a W boson is emitted. W and Z bosons are the force carriers for the weak force. Hopefully at this stage, the pieces will start to fit in.

The last step for total conservation is the lepton number. As discussed in the previous section, electrons are an example of a lepton. Now, we know that the initial neutron did not have any electrons in it, so when an electron is released from the W boson, an electron antineutrino has to be released as well. The electron has a lepton number of +1 (lepton number is different from charge!) and the electron antineutrino has a lepton number of -1. Together, this equals a lepton number of 0, and total mass and energy is conserved.

The exact same process occurs in beta plus, β^+, decay, just in the other direction. The proton changes into a neutron by converting one of its up quarks into a down quark, emitting a W boson. This time, instead of producing an electron and electron antineutrino, the opposite occurs. A positron and electron neutrino is produced instead, thus conserving the total mass and energy of the reaction.

To recap, the weak force allows neutrons to change into protons and vice versa during beta minus and beta plus decay, by changing the flavour of a quark from down to up and vice versa. In this process, the force carrier of the weak force, a W boson, is emitted. This W boson is then converted into an electron and electron antineutrino (or positron and electron neutrino in beta plus), to conserve the total mass and energy in this process.

So what exactly does this radioactivity do? The main process we have to remember here is the conversion of neutrons into protons and vice versa. The changing of the quark flavours. The process that occurs afterwards with the W boson and leptons can differ.

The weak force is useful in nuclear fusion, where it allows protons to change into neutrons and form deuterium (heavy hydrogen). This is one of the prerequisite steps for nuclear fusion, giving us light, heat, and chemical energy from the Sun.

Unstable isotopes, or radioisotopes, also decay through the proton-neutron conversion, providing many applications. The most prevalent among them is the use of carbon dating to estimate the age of fossils, rocks, and the Earth itself. It can also be used in cancer treatment and radiation therapy, or as tracers to detect problems in organs or tissues. Objects such as paper and pipes also owe their usefulness and effectiveness to the weak force. The thickness of paper and leakages in pipes can be regulated and detected using various radioisotopes.

The strong and weak nuclear forces also provide fascinating and endless possibilities for science fiction writers. *The Gods Themselves* by Isaac Asimov is one such example. In this story, two parallel universes are linked by the exchange of matter in what is called the 'electron pump'. Essentially, unstable elements are exchanged between the two universes, and due to the different physical laws, these elements undergo radioactivity and thus, produce energy. Another differing force is that the strong force in the parallel universe is ten times stronger than in our universe, meaning that they have smaller stars and face the threat of an ice age if their energy source runs out. It is a fascinating book, even telling the story from the perspective of the 'para-men', the aliens in the parallel universe. Do give it a read if you want to be awed by the wonders and consequences of these forces.

Ultimately, we have seen that the weak force is not 'weak' at all. Infact, the main reason it is dubbed 'weak' is due to its lower relative strength compared to the electromagnetic force and the strong force. It is simply 'weaker' than the other two forces. The weak force is actually studied in tandem with the electromagnetic force, to produce the electroweak theory.

We now emerge from the depths and confinement of the nucleus. Let us broaden our horizons ever so slightly and explore infinity. The weak force only makes up half of the electroweak theory. The other half is the familiar electromagnetic force. Worry not weary reader, it gets easier from here on regarding the four forces of our universe. The hard yards are behind us.

Imagine a game of baseball. The pitcher has a ball firmly in his hand. He throws it towards the batter. The batter, gripping the bat, swings and connects with the ball. The ball is impacted with the force from the bat and flies into the air. Eventually the ball is caught by a fielder and the batsman is out. All the basic, everyday phenomena mentioned here, such as holding, throwing, hitting, catching, even standing, are effects of the electromagnetic force.

The electromagnetic force acts between charged particles. Particles can either be positively or negatively changed, and we know that like charges repel and opposite charges attract. The closer and greater the charge, the stronger the force between the two oppositely charged objects. As the distance between the two objects increases, the force decreases. Theoretically, the force between the two objects can stretch into infinity, but the forces are so minute at great distances that it bears no consequence.

This is the same force which keeps electrons bound in an atom, as it is attracted to the positively charged proton in the nucleus. It prevents the negatively charged electrons from flying out of the atom. This force also allows for the formation of larger molecules and compounds due to the attraction between the oppositely charged particles. Without the electromagnetic force, the universe would just exist as a sea of atomic nuclei and free floating electrons.

The electromagnetic revolution was brought about in the 19th century. Many prominent scientists of the time worked tirelessly to achieve stellar breakthroughs in electricity and magnetism. The world has not forgotten the founding observations of current and compass needle by Ørsted, the equations of Maxwell, and Faraday's Law of Induction. It is said that the 19th century brought electricity to the masses while the 20th century brought the development of the gravitational and quantum world with achievements in infrastructure, medicine, technology, and the birth of space travel. What scientific revolution will the 21st century bring?

The electromagnetic force is carried by the photon elementary particle. This is the same photon from our discussion on the Cosmic Microwave Background. A photon is essentially a packet of electromagnetic radiation.

Let us return to our starting analogy of the baseball game. So how does the electromagnetic force allow for contact, feeling, pressure, friction, etc? The solution is in the mutual repulsion of the negatively charged electrons in the atom on each object. There is another reason called the Pauli exclusion principle which also prevents objects from passing through to each other but I won't go into detail here.

Take the interaction between the pitcher's hand and the ball for example. The reason the pitcher can feel and hold the ball is due to electrons in the atoms of the ball repelling the electrons in the atoms of the hand. This repulsion creates 'feeling' and the sense of touch. It allows the batter to hit the ball and it allows the fielder to catch the ball. So in reality, nothing is ever truly in contact with us. The contact we feel is just the repulsion of the electrons. Remember, the atom is 99.99% empty space. The reason we dont feel this empty space and why objects don't just pass through each other is due to this facet of the electromagnetic force. Even the normal force which pushes upwards on everything is due to the electromagnetic force.

You might be thinking that the electromagnetic force must be incredibly strong to support the weight of a book on a table, or the table on a floor, or even a human, and you would be right. The electromagnetic force has a relative strength of 10^{36}. Slightly weaker than the strong force, but stronger than the weak force. It seems rather simplistic when put like that.

Last and actually least, we arrive at gravity. Among the four forces, gravity is usually the odd one out. Not only is it considerably weaker than the other forces (1 : 10^{33} : 10^{36} : 10^{38}), it is also the only force without its own observed and detected elementary particle, although the existence of the graviton has been hypothesised.

The other three forces work their magic in the confinements of the atom. The strong force dictates proton - proton interactions, the weak force covers the neutron/proton conversion, the electromagnetic force allows for the attraction and repulsion of the subatomic particles. In terms of the atomic scale, gravity has virtually no part to play. However, when we move to larger, more grandiose scales, it is gravity that has the last laugh.

We must first understand the fundamental equation of gravity. Not Einstein's equations of relativity, but Newton's Law of Gravitation. This law refers to the force of attraction between two objects of some mass. The equation has been discussed in the previous chapter but we will cover it anyways, in slightly more detail.

$$F = G\frac{m_1 m_2}{r^2}$$

There are five factors to take into account;

1. The force of attraction between two objects, F.
2. The universal gravitational constant, G. This constant is given a value of $6.674 \times 10^{-11} \, Nm^2 \, kg^{-2}$.
3. The mass of the first object, m_1. This is usually the heavier of the two objects.
4. The mass of the second objects, m_2. This is usually the lighter of the two objects.
5. The distance between the centre of mass of the two objects, r^2.

Chemistry of the Universe

There are a few postulates that can be taken from this equation;

1. Every object in the universe exerts an attractive gravitational force on every other object. Whether it is a quark, an atom, a piece of chalk, a human, a planet, etc. If it has mass, it exerts gravity on anything else that has mass.

2. The magnitude of this gravitational force between two objects relies on the product of their mass. This explains why we do not feel the attractive gravitational force in everyday life between everyday objects, even though they are close together. The force is there, but the mass is so small that the force is almost negligible. This is also due to the gravitational constant having such a low value. For any perceivable force to be felt, it would have to at least counteract the gravitational constant, G.

3. The magnitude of force is also dependent on the distance between the centre of mass of the two objects. This is called the inverse square law, since the force decreases or increases based on the square of the distance, r^2. So no matter how far away two objects are, if they have mass then they exert some force of gravity on each other. Theoretically, the Earth has attractive properties with everything in the solar system. The entire universe for that matter. However, given the vast distances of space and the Earth not having much mass anyways in the grand scheme of things, all we can attract is the Moon (and a few stray asteroids of course).

In short, the heavier two objects are, and the closer they are to each other, the stronger the force of gravity. This clearly explains why gravity has little place in the world of atoms, but rules supreme in the land of black holes and neutron stars.

Exploring the Essence of Everything

Although, we do have to remember those initial density fluctuations post Big Bang which gave rise to all the structure in the universe. Whole galaxies born from miniscule fluctuations of gravity where the force wouldn't have been more than the force between the atoms in our body. And yet, here we are. The product of 13.8 billion years of gravitational evolution.

Be that as it may, the binding of quantum mechanics and gravity still poses a daunting challenge to scientists. How can we envelope these two seemingly distant theories into one Theory of Everything? The world of the micro and the world of the macro? We can more or less understand how the other three forces are connected in the Grand Unified Theory, but how does gravity fit in? What was its role during pre-Big Bang conditions? Many questions to be asked, and even more solutions to be found.

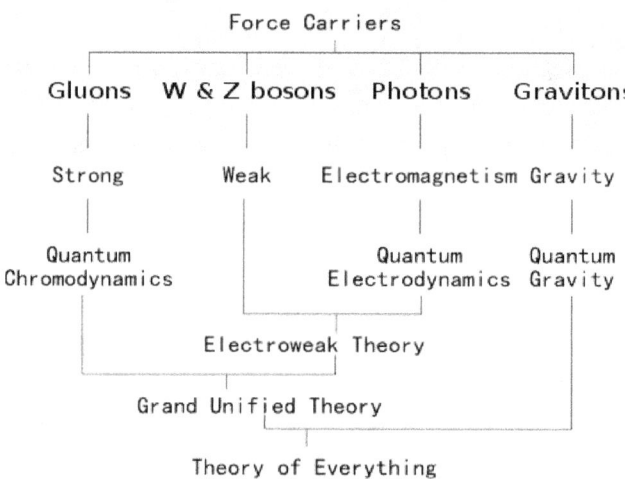

You may have noticed that I have omitted the history and scientists involved in the discovery and development of the Standard Model and the fundamental forces. This is mainly due to the broad range of content already provided, and the many layers of history and scholars involved.

However, I believe that these pioneers shouldn't be left out completely, so a summary on a few individuals involved in the experimentation and development of theories on these particles and forces is provided in the gallery at the end of the chapter.

Speaking of the Big Bang, we can finally complete our dissection of the phases during the early evolution of our universe. We last had a primordial soup of the 'quark-gluon plasma'. We now know what quarks, gluons, baryons, and all the other forces are. However, elementary matter particles were not the only characters in play during this time. We once had an entirely different set of particles, and the entire evolution of the universe could have been oh so different. Where there is matter, there was also antimatter.

Angels and Demons

In Dan Brown's best selling novel *Angels & Demons*, (as well as in the film adaptation starring Tom Hanks), the plot revolves around antimatter, bombs, Vatican secrets, popes, and the Illuminati. It is perhaps the most famous work in pop culture to represent the workings of antimatter on such a scale.

In the story, a physicist at CERN has discovered how to create antimatter. Shortly after this discovery, the physicist is murdered by members of the Illuminati and the sample of antimatter stolen. Professor Robert Langdon (a recurring protagonist in Brown's subsequent books) is tasked to solve the mystery and to recover the antimatter which will explode in the Vatican City, in 24 hours. The stakes are set for a hectic and breakneck day.

Obviously I will not reveal any intricate details of the plot here, feel free to watch the movie or read the book, but it does share an interesting albeit ominous idea. The prospect of not just harnessing antimatter, but using it as an energy source. However, this comes with many profound dangers and difficulties.

Exploring the Essence of Everything

Firstly, let's start at the basics. What is antimatter and why is it studied so strenuously? Antimatter is matter consisting of antiparticles. We have already seen examples of antiparticles in beta decay, namely the electron antineutrino and the positron. Each particle of matter has its own antiparticle; the antiparticle of the electron is the positron, the antiparticle of the proton is the antiproton, the antiparticle of the quarks are antiquarks and so on. The matter/antimatter pairings have the same mass, but opposite physical charges. For example, the positron is positively charged while the electron is negatively charged. Some particles, such as photons, are their own antiparticle.

Antiparticles can form antimatter the same way ordinary matter does. An antiproton of negative charge can bond to a positively charged positron to form an antihydrogen atom. In theory, an entire universe identical to ours could be formed from antimatter, with all its charges reversed. But if there is the possibility of forming antimatter from antiparticles, why don't we see antimatter in our own universe? Why is our universe dominated by regular, ordinary matter? Why is antimatter so rare and hard to produce?

To answer these questions, we have to go back to our Big Bang phases and probe Einstein's famous equation, $E = mc^2$. In a nutshell, this equation states that any amount of mass can be converted into pure energy and vice versa. The higher the mass, the more energy you obtain.

Using Einstein's equation, we can substitute factors in for energy and mass. Our energy is carried by the photon, and mass is expressed by matter/antimatter. A single photon with sufficient energy at high enough temperatures can create a single matter/antimatter pair, which will then annihilate, creating the photon again. Imagine the energy required to produce such a pairing. This exact same energy is released when the two particles touch and annihilate. This is the concept used in Dan Brown's *Angels & Demons* to produce an antimatter bomb. The cycle repeats over and over, instantaneously. But it was only prevalent in a specific period of time.

Chemistry of the Universe

It is now 10^{-15} seconds after the Big Bang. This is the start of our 'photon to matter/antimatter back to photon' cycle. Temperatures are hot enough for photons to convert their energy to pure mass and the subsequent annihilation back to photons. This cycle occurred for every single particle/antiparticle pairing. The electron/positron, the quark/antiquark, the boson/anti boson, etc.

A few fractions of a second later, at 10^{-11} seconds, the baryons were formed (particles consisting of three quarks). As the baryons were formed, so were the antibaryons. More annihilations preceded. Baryon/antibaryon parings would pop in and out of existence only to annihilate and restart the cycle over and over again (all this in a fraction of a fraction of a second).

Finally, at 10^{-5} seconds, the universe has sufficiently cooled so that photons no longer have enough energy to form matter/antimatter pairings. The final annihilations are now taking place. But then this raises the question, why is there matter at all in our universe? Shouldn't it all have self annihilated along with antimatter during this period? Why isn't our universe just composed of photons? A universe of pure energy with no matter. This is one of the greatest mysteries in astrophysics. The asymmetry and imbalance of matter and antimatter.

During this period of matter/antimatter annihilation, there was some residue. For every one billion annihilations, there would be one particle of matter left over. This is the asymmetry of matter and antimatter. A billion and one to a billion. But why there is one extra matter particle, scientists are none the wiser. Naturally there are theories and experiments conducted in particle accelerators, but a definite explanation remains unattainable for now.

So one matter particle for every billion annihilations survived. The mystery does not stop there. What is also bizarre is that this asymmetry applies to every type of particle. Not just quarks, or leptons, or protons. Every single type of particle had this billion to one asymmetry. What are the chances of that?

Exploring the Essence of Everything

After the final matter/antimatter annihilations, all that was left was the accumulation of these single matter particles (and the photons of course). These particles formed everything we see in our universe today. If not for this asymmetry, there would be no single particles leftover, there would be no matter whatsoever in our universe. Just a never ending sea of pure energy photons. This is why antimatter is so rare in our universe. They were all annihilated during the first second of the universe's existence. But that doesn't mean that they're out of the picture forever.

The modern theory of antimatter began in 1928, with a paper written by English physicist Paul Dirac. Dirac realised that the Schrödinger wave equation for electrons predicted the possibility of antielectrons (or postirons). As with most discoveries in physics, the theory came before the practical experimentation. The actual physical discovery of antimatter was first done by Carl David Anderson. In 1932, he conducted research on cosmic rays and encountered unexpected particle tracks in his photographs. These particle tracks were similar to that of an electron in mass, but had the opposite electrical charge. This discovery validated the theory made by Paul Dirac in his paper, and Anderson shared the 1936 Nobel Prize in physics along with Victor Hess, who had discovered cosmic rays in the first place.

In the present day, antimatter can be created either naturally or artificially. Naturally, it is emitted in beta decay, produced in cosmic rays, and can be found in the jets of neutron stars and black holes. As long as the temperature is high enough, energy can be converted into particle/antiparticle pairs akin to Big Bang conditions.

However, most of the antimatter magic occurs right here on Earth. When physicists discuss antimatter, there are usually four topics of discussion; production, preservation, cost, and function.

Production of antimatter occurs in particle accelerators, mainly at CERN. Usually, a beam of protons is fired at a block of metal, and the resulting particle debris from the collision is studied. This debris can consist of our elementary particles, as well as a handful of antiprotons.

These fragments of antimatter exist for very brief periods of time before annihilating. This leads us to the next topic, preservation. How can we possibly trap these antiparticles long enough to study them?

In truth, most of the 'studying' and analysing is done in the post-mortem of these collisions. The incredibly dense amounts of data being fed into supercomputers per second will be picked apart and sorted by scientists who will then try to identify signatures of antimatter. To actually preserve and trap antimatter we have to use electric and magnetic fields, and a vacuum. Of course, using any form of ordinary matter such as a glass container or even interactions with air for that matter will cause an annihilation.

Fear not, the amounts of antimatter being formed in these experiments are so tiny that any annihilation will yield inconsequential amounts of energy. After all, the energy produced depends on the mass of the system. However, get enough of the stuff and the outcome could be disastrous. Even 1 gram of antimatter could produce an explosion rivalling an atomic bomb. To put that into perspective, at the present day we have only made around 15 nanograms of the stuff. Perhaps you could use it to light a match.

In 2011, scientists at CERN were able to preserve a sample of antihydrogen for up to 17 minutes. The current record (also at CERN) is storing a sample of antiprotons for 405 days.

Next, the taxpayers and politicians headache, and every scientist's primary stumbling block; how much is it going to cost? There are no two ways about it. Producing and maintaining the temperatures and conditions to create antimatter is expensive. Building a particle accelerator is even more expensive. So expensive in fact, that antimatter is the costliest substance to make. According to CERN, it has cost a few hundred million Swiss Francs just to produce 1 nanogram of antimatter. In 1999, NASA gave an estimate of 62.5 trillion dollars for a single gram of antihydrogen. Clearly, a little goes a long, long way and a lot produces a small, small amount, when it comes to antimatter. So why bother? What are the practical uses of antimatter?

Beta decay and radiation has many applications in the field of medicine. Positron Emission Tomography (PET scans) relies on short lived isotopes which act as tracers when introduced in the body. This helps in the detection of tumours and the imaging of organs in the body.

The main use of antimatter is in terms of energy. For good or evil, for angels or for demons, we cannot shun the idea that we can obtain such vast amounts of energy from such miniscule amounts of a substance. Manufacturing, preserving, and cost is another matter. The benefits on their own should be discussed regardless.

In terms of energy, we have to explore the idea of space travel using antimatter fuel in engines. Since antimatter can store more energy with less mass, the thrust to weight ratio of the spaceship would be much higher and efficient compared to a regular spaceship. In short, we obtain much more energy, for a fraction of the mass, in turn reducing the amount of energy we need, and increasing the efficiency of the entire operation. To organise an operation on this scale however, antimatter would have to be made at industrial levels.

So ends our discussion on the mystique and wonders of antimatter. From bombs in pop culture, to the costliest projects ever undertaken, to the potential for interstellar travel, and to the very existence of our universe today. Watch the antimatter space carefully. There is more to this hidden world of particles than meets the eye.

In the meantime, let's shift our gaze back to the ordinary, mundane matter of our universe. Where do we go from four forces and seventeen fundamental particles? How were the first atoms and elements in our universe created?

Genesis of the Elements

The next three sections will cover the formation and creation of the elements. From simple hydrogen with its single proton and electron to giants of the elemental words, all will be explained and revealed.

Chemistry of the Universe

The creation of new atomic nuclei from preexisting nuclei is called nucleosynthesis. There are three types of nucleosynthesis; Big Bang nucleosynthesis, stellar nucleosynthesis, and supernovae nucleosynthesis. This section will cover Big Bang nucleosynthesis.

One thing to clear up before we begin, the formation of new nuclei is not the same as the formation of atoms of the elements. Atoms contain protons, neutrons, and electrons, whereas the nuclei just contain protons and neutrons. Electrons were not bonded to nuclei until the universe had sufficiently cooled, some 380,000 years after the Big Bang. This is the exact same time when photons stopped crashing into all those rogue electrons and the CMB was released.

So while the formation of the first atoms only occurred hundreds of thousands of years after the Big Bag, nuclei formation happened almost immediately, and ended abruptly. A period of intense nuclear fusion, starting 2 minutes after the Big Bang, and ending 18 minutes later.

It is also crucial to remember that the pure hydrogen nucleus contains just a single proton. There are no neutrons in hydrogen-1. So once hadrons (which include protons) were formed through interactions of the strong force, it can be said that the universe was populated with hydrogen-1.

It is also important to specify the different types, or isotopes, of hydrogen. An isotope is an atom of an element (or in this case just the nuclei), that has the same number of protons but different number of neutrons. This will be familiar to any of you that have taken a high school chemistry class. Hydrogen has two isotopes, deuterium and tritium. For simplicity's sake, pure hydrogen with just 1 proton and 0 neutrons will be referred to as 'hydrogen'. The isotopes of hydrogen will be referred to by name, instead of hydrogen-2 or hydrogen-3.

Another familiar concept is that all elements are differentiated by their proton number. Of course, there are a whole range of differences between elements, but the key difference is the number of protons each element has.

Alright back to the Big Bang. We now have a population of single, lone protons floating in the void of space. The next step is to bond these single protons to the single neutrons. Out of all the protons, only 25% attach to neutrons to form heavy hydrogen, deuterium. In fact, deuterium is the main reason why Big Bang nucleosynthesis started and ended so quickly. Let me explain.

Immediately after deuterium is formed, the nuclei will form helium-4 (in reality there are a few intermediate steps but helium-4 is the end product that we will focus on). So we have gone from hydrogen to helium through deuterium. The key is that to form helium we have to first form deuterium. But deuterium is not as easy to make as first thought.

The formation of deuterium is mediated by the strong force which holds the single proton and neutron together. This strong force is also affected by temperature, especially the fluctuating temperatures post Big Bang. Before 2 minutes, temperatures were too high and any deuterium that forms would easily break apart. After 2 minutes, temperatures fall and deuterium becomes stable, albeit briefly. From there, helium-4 is able to be formed.

However, once that timer reached 20 minutes, the temperature was too low due to the expanding universe, and was no longer sufficient to fuse hydrogen together to form deuterium, and hence, helium. Remember, this process of fusion usually happens in the core of stars, at millions of degrees.

During this short time period, only three types of elemental nuclei were formed. Hydrogen, helium, and lithium. The formation of elements heavier than helium was severely hampered due to the short time period of fusion, and the falling temperature and low density. Remember from the previous chapter that it takes stars thousands to million of years to form these heavier elements, under severe temperatures and densities. Thus, only trace amounts of lithium were formed.

The idea of Big Bang nucleosynthesis was first brought to attention by Ralph Alpher in the 1940s, while he was a phD student at the George

Washington University. Alpher, along with his advisor, Russian physicist George Gamow, argued that the Big Bang would create hydrogen, helium, and heavier nuclei to explain their abundance and composition in the early universe.

The paper in which the theory was published was called the Alpher-Bethe-Gamma paper, or αβγ paper, as in the three forms of radiation. This was achieved by the inclusion of German physicist Hans Bethe who initially had nothing to do with it whatsoever. Gamow included Bethe's name to produce the so-called 'Alpha-Beta-Gamma' moniker, but Bethe did work on Big Bang nucleosynthesis after it was published.

The αβγ paper successfully explains how the first nuclei were formed from post Big Bang conditions, and their abrupt halt due to decreasing temperatures and density. It also explains the abundance and composition of the nuclei in the early universe.

So to recap, the first few stable nuclei were hydrogen, deuterium, helium, and lithium, formed through nuclear fusion when the universe was still dense and hot. This only lasted until 20 minutes post Big Bang, when conditions become unfavourable for fusion. The composition would have been 90% hydrogen to 10% helium with deuterium and lithium existing in trace amounts. This composition stayed unchanged until the formation of the first stars, where heavier elements were introduced. That is where we venture next.

In the Heart of a Star

Even though our Sun is 4.6 billion years old, the mystery as to how it derives its energy to produce light, heat, and elements, remained unsolved until just over 100 years ago. It was the English physicist and astronomer Arthur Eddington, in the 1920s, who first proposed that stars obtain their energy from the process of nuclear fusion, turning hydrogen into helium. Following that proposal, a number of scientists continued research into this area of nucleosynthesis, among them George Gamow, Hans Bethe, and Fred Hoyle.

George Gamow derived the so-called 'Gamow factor' which is used to determine the probability that two nuclei will overcome the Coulomb barrier between them and undergo nuclear fusion. But what is this Coulomb barrier?

In our atom, we have the electrostatic force between the electron and proton, as well as the strong force between the protons in the nuclei. The repulsion of two positively charged particles prevents them from coming too close together. This is called the Coulomb barrier. For two protons or positively charged nuclei to come close enough for the strong force to take effect and undergo subsequent nuclear fusion, the nuclei have to overcome this Coulomb barrier. Basically overcome the repulsive force preventing it from getting too close to the other nuclei. In order to do this, the nuclei will require high velocities, high temperatures, and high densities. These are the conditions for nuclear fusion.

Once Gamow had obtained this probability of two nuclei overcoming the Coulomb barrier, Hans Bethe analysed the different ways in which hydrogen could turn into helium. He considered two separate processes, the proton-proton chain reaction, and the carbon-nitrogen-oxygen cycle. We will go through the details of each cycle in a bit. However, Bethe's proposal only considered fusion reactions until oxygen, but we know that heavy stars produce elements until iron.

It was Fred Hoyle who finally completed the picture of stellar nucleosynthesis with a theory in 1946 and a paper in 1954 detailing how heavy, massive stars could synthesise elements from carbon to iron.

Let's return to our chronology of the universe and see where we stand. 20 minutes after the Big Bang and the universe is permeated by hydrogen, helium, and trace amounts of lithium. The next 100 to 200 million years are relatively uneventful, in terms of nucleosynthesis. Big Bang nucleosynthesis has come and gone, and stellar nucleosynthesis won't occur until the first stars form.

Chemistry of the Universe

Once the first stars form, some 100 to 200 million years after the Big Bang, stellar nucleosynthesis kicks in. The fusion of hydrogen occurs differently, depending on the mass and size of the star. For smaller, average sized stars such as the Sun, the dominant method is the proton-proton chain reaction. For larger, heavier stars, the dominant reaction is the carbon - nitrogen - oxygen cycle (CNO cycle).

The proton-proton reaction is similar to Big Bang nucleosynthesis in that it creates deuterium as an intermediary product in the complete reaction. Refer to the diagram on the right for the complete process. In the reaction, neutrinos, gamma rays, and positrons are emitted. In fact, the Sun produces 10^{38} neutrinos per second. There are trillions of neutrinos passing through you at this very moment! Don't be alarmed, neutrinos barely interact with matter, and travel at light speed, which makes them extremely difficult to detect and measure.

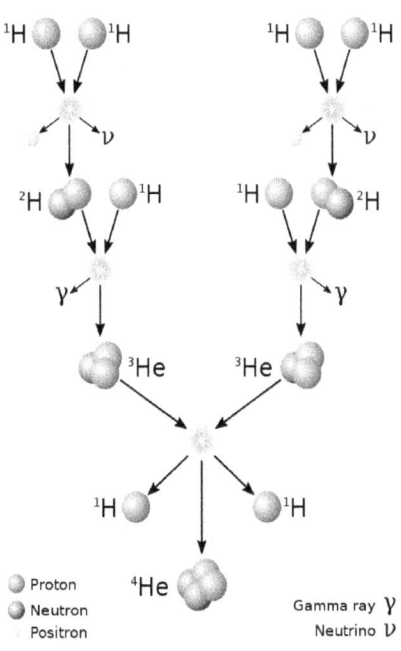

Well what about the gamma rays? Surely the amount of gamma rays produced by fusion reactions will be detrimental to our health? We have to remember that the conditions in the core of stars undergoing fusion are extremely dense with hydrogen atoms. The gamma rays that are emitted constantly collide and interact with these atoms, losing energy in the process. Once the rays reach the outer surface of the Sun, most of the energy has been lost and the rays exist as either UV, visible light, or infrared. Infrared rays are responsible for the heat we feel, visible light allows us to see, and the majority of UV rays are absorbed by the ozone layer.

These collisions are so extreme and frequent, that it can take photons (gamma rays essentially) up to 150,000 years to travel from the core to the corona! Once it reaches the corona and is emitted, it only takes 8 minutes to reach Earth, due to the speed of light. So just imagine. The light and heat that you are experiencing right now, actually originated in the heart of the Sun at the same time that the very first humans were emerging on the Earth. The fusion reactions happening at this moment might not even be experienced by civilization, depending on the state of the world in 150,000 years.

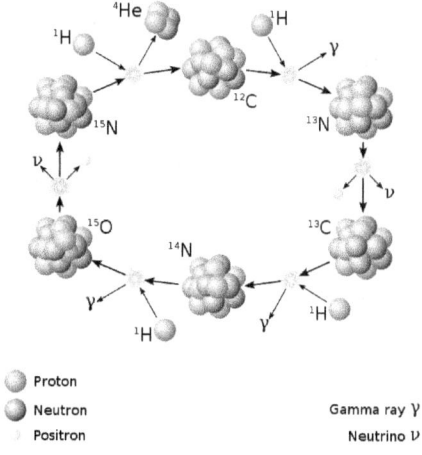

In comparison to the proton-proton chain reaction, the CNO cycle is much more complex. This is due to the use of carbon, nitrogen, and oxygen as intermediaries in the cycle. Just like the P-P cycle, the CNO cycle also emits gamma rays, positrons, and neutrinos. However, the CNO cycle requires much higher temperatures to self-maintain compared to the P-P cycle, and thus is mainly used in stars of 1.3 solar masses and above.

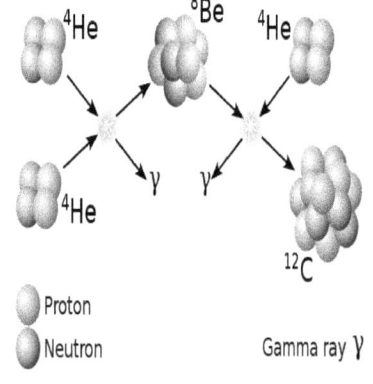

The next step is helium fusion. Helium is fused to carbon via the triple alpha process, in both large stars and smaller Sun-like stars. The triple alpha process fuses two helium-4 nuclei to form an intermediary beryllium-8 nucleus. Another helium-4 fuses with the beryllium-8 to produce carbon-12. This process only emits gamma rays. The reason this is called the 'triple alpha' process is due to the three helium-4 nuclei involved. An alpha particle is essentially a helium atom that has been stripped of its electrons, and consists of 2 protons and 2 neutrons. 3 of these alpha particles will form a carbon-12 nucleus.

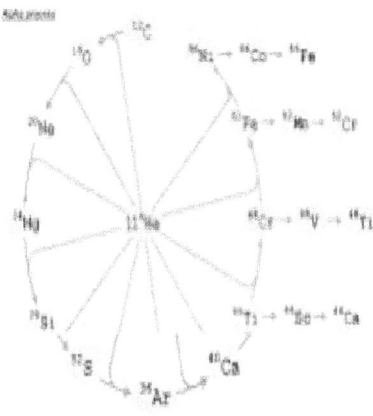

We know from the previous chapter that Sun-like stars stop their fusion at carbon and oxygen. Some stars will stop at the triple alpha process, and some will carry on to form oxygen in the first step of the alpha process. The alpha process is what allows massive, heavy stars to form the elements up to iron.

The alpha process essentially adds more and more helium-4 nuclei (hence the name) to each element to produce heavier elements. The first step is adding a helium-4 nucleus to a carbon-12 nucleus. This produces an oxygen-16 nucleus. For some Sun-like stars, their entire process of fusion stops here. For the others, more and more helium-4 is added to produce neon-20, magnesium-24, silicon-28, sulphur-32, and so on, until iron-56 is reached. As usual, gamma rays are emitted during every fusion reaction.

To understand why fusion stops at iron, we have to understand the difference between an exothermic process and an endothermic process. In a chemical reaction, total energy is either lost or gained. An exothermic process will lose more energy to the surroundings than it gains, whereas an endothermic process will gain more energy from the surroundings than it loses.

In a nuclear fusion reaction, energy is supplied by the high temperature in the core of the star, while energy is lost, or emitted, during the fusion process itself. For elements lighter than iron, more energy is released than is absorbed, resulting in a reaction that can sustain itself. The energy that is lost to the surroundings is then re-used in the next fusion reaction and so on. The remaining energy is lost to the surroundings as heat and light.

For elements heavier than iron, the process of fusion is endothermic. This means that more energy is required to carry out fusion, due to the stability of the iron nuclei, than energy that is obtained from the reaction. As a result, the fusion reactions are not sustainable anymore, since more and more heat is being used up, decreasing the overall temperature of the core until fusion is no longer possible.

We now know how to obtain hydrogen, helium, and lithium nuclei from the primordial soup of elementary particles in Big Bang nucleosynthesis, and how elements in a star are manufactured, from helium all the way to iron. However, we have only covered around half of the naturally occurring elements in our universe. To find out how the remaining half are made, we turn to the final stage of fusion, supernovae nucleosynthesis.

Life From Death

We know from the previous chapter that larger stars reach the end of their lives once iron is formed in their cores, and then they have a violent death in the form of a supernova. This supernova is crucial for two reasons. One, it releases the heavy elements manufactured in the core of the star. These elements get ejected into interstellar space, for new stars and planets to be formed. Some of these planets might even harbour life. We see that without supernovae, elements would be trapped inside stars, and life would never have been possible. Two, the supernova itself manufactures elements heavier than iron. I glazed over this process in the last chapter, but we will explore it in greater depth now.

The importance of supernovae was first put forth by our familiar friend Fred Hoyle. We know that he was a key player in stellar nucleosynthesis, but he also contributed greatly to supernovae nucleosynthesis.

In 1957, Fred Hoyle co-authored a paper with scientists Eleanor Burbidge, Geoffrey Burbidge, and William Fowler. It became known as the B^2FH paper due to the initials of the four authors. This paper spoke about a phenomenon that occurs in a supernova, where the neutron-rich isotopes of elements heavier than nickel can rapidly capture freely moving neutrons. This process is called the r-process, short for rapid neutron-capture process. Following the publication of the B^2FH paper, experiments were carried out to test this theory and it was subsequently proven correct.

As special as the r-process is, it only accounts for half of the heavy elements made in supernovae. The B^2FH paper also spoke about a second process in which the other half of heavy elements are manufactured. This process is much slower, spanning thousands of years, compared to the microseconds of the r-process. As expected, this is called the s-process (slow neutron-capture process).

There is a unique feature of the elements that further proves the difference between stellar and supernovae nucleosynthesis. Each element (barring pure hydrogen) has a proton to neutron ratio in its nucleus. For every element, the number of protons is fixed while the number of neutrons can differ. These are known as isotopes of the element. When we examine the proton-neutron ratio of the elements below iron, we see that they are almost one to one. Carbon-12 has 6 protons and 6 neutrons, oxygen-16 has 8 protons and 8 neutrons, magnesium-24 has 12 protons and 12 neutrons, and so on. This one to one ratio causes these atoms to be very stable. As this ratio gets more and more uneven, the atom becomes increasingly unstable.

The change in the proton-neutron ratio occurs in these neutron capture events. The key difference between nuclear fusion and neutron capture is that since neutrons are electrically neutral, it is much easier for them to be added into a nucleus compared to the positively charged proton. Thus, the proton-neutron ratio becomes more uneven as more neutrons are bonded to the nucleus. This results in the formation of highly unstable radioactive nuclei. So how does an atom lose neutrons while gaining protons to shift this ratio back to stability?

The answer is in beta decay. We have seen that beta decay can change a neutron into a proton plus an electron. This causes the number of neutrons in the nucleus to decrease by one, while the proton number increases by one. As the proton number changes, so does the type of element. This produces a new, more stable element.

The r-process doesn't just occur in supernovae, it can also happen during neutron star mergers. This field of research is relatively new in terms of obtaining feasible results, as it requires the study of gravitational waves as well as cooperation between a number of observatories and detectors worldwide such as LIGO, VIRGO, and INTEGRAL.

These three processes of nucleosynthesis (Big Bang, stellar, supernovae) only account for 92 of the 118 elements in our universe. These 92 are called the naturally occurring elements. The remaining 26, the transuranium elements, are man made in laboratories and are named after notable scientists. For example, fermium, curium, einsteinium, and rutherfordium.

Alas, we reach the end of our journey on the production of elements in our universe. From the initial duo of hydrogen and helium 2 minutes after the Big Bang (a trio if you include lithium), to the stellar processes producing carbon, oxygen, and iron to name a few, to the massive explosions birthing new elements such as gold, uranium, and silver, to the final artificial elements made much closer to home.

Our brief foray into beta decay and alpha particles brings up another (and frankly quite important) feature of elements and atoms:

radioactivity. We know what these atoms are made from, we know how to produce them, but how do we break them up? How do we transmute them at will? And how might we use this ominous process to the benefit of humankind?

Making Superheroes and Supervillains

Radioactivity is a curious topic, pun absolutely intended. At first glance, most people shun it. Perhaps that is the current stigma. But how can we blame them? Thoughts of Chernobyl and scenes of atom bombs falling from the sky are shrouded in radioactive fallout and agonising death. On the flip side, we use radioactivity to cure our most dangerous diseases, and to determine the age of fossils and the Earth itself. So where do we draw the line from outright mutilation by radiation and the benefits of a few cosmic rays?

Over the past few decades, science fiction writers, especially comic book writers (the two are not mutually exclusive), have delved into this realm of radiation to create the most famous superheroes and supervillains of our time. In fact, it is a rather ingenious way to explain how a teenage boy can crawl on walls, how a man can manipulate the electromagnetic spectrum, and the existence of a fantastic group of individuals who charted in the 60s (the comic book nerd in you might understand that reference).

How do gamma rays turn Bruce Banner into a mean green smashing machine? How do cosmic rays turn one person to stone and another to a human fireball? How does radioactive toxic waste cause a blind lawyer to fight crime on the side? These are not simply works of fiction. In some ways, every one of these outcomes is technically possible.

I'm not saying that we should submit ourselves to the full extent of gamma radiation in order to turn into the Hulk or throw ourselves into a nest of spiders to become Spider-Man, but this type of imagination from writers shows us the possibilities of radioactivity.

Exploring the Essence of Everything

Before becoming a staple in comic books, radioactivity had to become a staple in the laboratory. And this was first achieved at the end of the 19th century, with the works of two scientists; Wilhelm Röntgen and Henri Becquerel. In articles on radioactivity, Röntgen's involvement is usually omitted but his work really got the ball rolling, or in this case, got the rays glowing.

In 1895, Röntgen conducted an experiment using a light bulb, a piece of black cardboard, an induction coil, and a screen of barium platinocyanide. He covered the bulb with the black cardboard to prevent any light from escaping, and attached it to the induction coil. He then noticed a faint glow on the barium platinocyanide screen, placed a few feet away. How could the light pass through the black cardboard and create a glow on the screen? This experiment marked the discovery of X-rays, but that is another story.

A year later, Beqeurel investigated this 'glow in the dark' property of certain materials. These materials glowed after being exposed to light, but Bequerel wanted to pinpoint which exact materials glowed and why. He tried a number of phosphorescent salts, all of them failing to achieve the desired outcome. He finally struck gold when he used uranium salt. This uranium substance managed to glow on its own accord!

Initially, he believed that these 'uranium rays' were similar to the X-rays discovered by Röntgen a year prior. However, further research from a group of scientists, including the famed couple, Pierre and Marie Curie, proved otherwise.

Marie Curie used an electrometer (a device developed by Pierre and his brother to measure electric current) and discovered that a uranium sample caused the air around it to become charged. She also found out that the activity of the uranium only depended on the quantity present. More uranium would cause more electrical conductivity of air and vice versa. Using these two observations, she hypothesised that the radiation was not the outcome of an interaction between molecules, but must originate from the atom itself.

Pierre and Marie continued to isolate different materials and elements that showed signs of radioactivity, and by doing this, they found two completely new elements; polonium and radium.

For all these achievements, there were awards to be given and accomplishments to recognize. Starting in 1901, Wilhelm Röntgen received the Nobel Prize for his discovery of X-rays. A couple years later, the trio of Henri Becquerel, Pierre Curie, and Marie Curie received the Nobel Prize in physics for their research on radioactivity. In 1911, Marie Curie became the first person ever to receive two Nobel Prizes in different fields, when she received the Nobel for chemistry due to her discovery of radium and polonium.

Unfortunately, Pierre was killed in a road accident 5 years earlier and did not share in this award. In fact, the Curie family was plagued with untimely deaths due to their very own research. Marie, her daughter Iréne, and Iréne's husband, Frèdèric, all passed away due to the effects of radiation from their years of research in an unprotected, unsafe laboratory.

Once again we ask the question; how does this radiation cause so much damage and yet is used in hospitals to treat the most serious of illnesses? To understand this, we have to study the core of radioactivity.

Radioactivity is the process in which an unstable atomic nucleus loses energy. It is said to 'radiate' this energy, thus giving rise to the term. A material containing these unstable nuclei is said to be 'radioactive'. The three primary forms of radiation are alpha decay, beta decay, and gamma decay. In truth, there are around 25 different ways in which an atomic nucleus can decay, but we will focus on the main three.

Just before the Curies conducted their experiments on uranium, radium, and polonium, the New Zealand physicist Ernest Rutherford had separated radiation into three distinct types; alpha, beta, and gamma. However, experiments and confirmation of theories on these forms of radiation were only carried out a few years later, at the dawn of the 20th century.

Exploring the Essence of Everything

In 1908, Rutherford's students, Hans Geiger and Ernest Marsden, conducted his famous 'gold foil test'. It had been proved by Rutherford himself a year previous, that alpha particles were identical to helium nuclei. They both consisted of two protons and two neutrons. The alpha particle was known to be positively charged due to a lack of electrons. Geiger and Marsden directed a beam of these alpha particles through a gold foil of only a few hundred atoms thick. What they observed set the stage for the entire development and structure of the atomic model.

The observed model of the atom at the time was J.J. Thomson's, 'plum pudding model' where the atom was a positively charged area of some volume (the pudding), with the negatively charged electrons embedded in it (the plums). However, when the alpha particles were directed at the gold foil, it was observed that most of them passed through without any interference! A handful were deflected at angles, and only a very tiny fraction bounced straight back. Clearly, the model of the atom had to change.

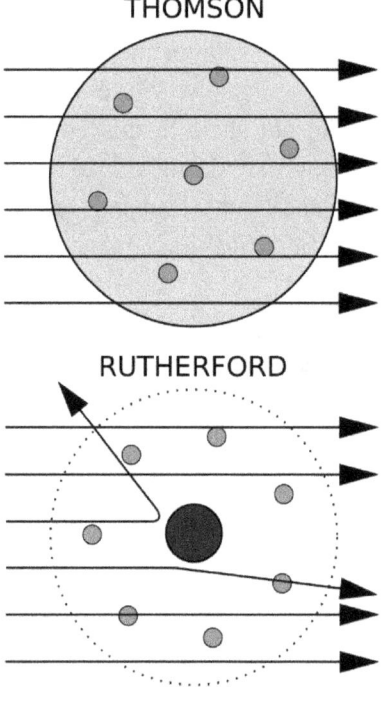

The majority of alpha particles passing through the gold foil meant that the gold atoms had to be mostly empty space. The deflection of the particles was due to the basic law of electromagnetism; like charges repel. Using this, Rutherford could determine that all the positive charge in an atom was concentrated in a tiny volume in its centre, and the negatively charged electrons orbited this centre. The alpha particles that passed close to the nuclei got deflected, and the few that hit the centre straight on, bounced back. Using these deductions, along with a few quantum principles of the electrons, Niels Bohr was able to finally determine the complete atomic model.

Beta decay has already been discussed in depth, regarding the weak force and decay in nucleosynthesis. As a result, we shall skip the intrinsic properties of this decay and move on to gamma rays.

Gamma rays are basically very energetic photons. You would have learnt in your high school physics class about the electromagnetic spectrum. Gamma rays sit right on the top of the spectrum, meaning that they have the highest energy, highest frequency, and shortest wavelength. Remember, the electromagnetic spectrum, or waves, are carried by photons.

The energy released in any fusion or fission reaction is released mainly in the form of gamma rays. Fusion in the Sun's core, radioactive decay of uranium, reactions in nuclear reactors. Even lighting strikes can produce gamma rays. Gamma rays rarely occur on their own. They are a by-produce of the energy lost in a reaction or decay. Almost every alpha and beta decay produces gamma rays as the leftover energy.

To get a complete grasp on these three forms of radiation, let us compare and contrast them. We will discuss two different properties; penetrating power and ionising power.

Penetrating power is how far the emitted ray can travel before being stopped, and how easy it is to block. This distance is determined by the mass, and therefore speed, of the particle emitted. Alpha radiation emits a helium nucleus with 2 protons and 2 neutrons, beta decay emits either an electron or a positron (depending on whether it is beta plus or beta minus), and gamma rays emit a photon. Of course, the alpha particles are the heaviest, and therefore travel the slowest and are easiest to block. The electrons and positrons of beta decay sit in the middle, much smaller than alpha particles but also much slower than photons. And the photons themselves are practically massless and travel at light speed.

Alpha rays have such low penetrating power that they can normally be blocked by a single sheet of paper. Beta particles usually stop at aluminium shielding. Gamma rays however, require much more effort to stop, due to their nonexistent mass and rapid velocity. It usually takes a thick layer of lead or concrete to shield from gamma rays.

There is a flip side to this law, which is ionising power. To ionise is to cause an atom to become charged. This is caused by either the removal or acceptance of electrons. In this case, electrons are removed when these radiation particles collide with the atoms, causing them to be ionised. Where penetrating power was inversely proportional to mass, ionising power is directly proportional to the mass of the particles.

In this case, alpha particles with their high mass have the highest ionising power. This high and dangerous ionising power is counteracted by its low penetrating power and thus they are relatively harmless to us since they cant pass through skin and tissue. Again, beta particles sit in the middle, having more ionising power than gamma rays, but less than alpha particles. Gamma rays have very low ionising power due to them having virtually no mass and no charge as well. However, remember that gamma rays have very high energies and are rather hard to stop, meaning that they can easily pass through our skin and tissues, harming our organs, and especially DNA. Even though the ionising power of gamma rays is low, the structure of DNA in our body is so small that it is easy for gamma rays to manipulate and alter these codes at will. This genetic mutation is the cause of most superhero and supervillain abilities in comic books.

If we are to speak of the dangers of radiation, then we must also say something about its benefits and uses.

Due to its low penetrating power but high ionising power, alpha particles are suitable to be used in everyday situations, and are generally healthier than gamma rays in medical treatment. Americium-241, is used in smoke detectors by emitting alpha particles. These particles ionise the air in the detector to produce a small current. When smoke fills this chamber, the current is reduced and the alarm is triggered.

Radium-223 emits alpha particles during treatment of bone cancer to try and kill the malignant cancer cells. Alpha particles can even power space probes and pacemakers. This is done by using a device called a radioisotope thermoelectric generator which converts the heat energy released in the decay of these radioisotopes into electrical energy.

Beta particles are also used to treat cancers. Radioisotopes such as lutetium-177 and yttrium-90 are ingested into the body and once inside, decay to produce beta particles which kill cancer cells. The dosage of these particles is controlled to prevent any adverse effects such as genetic mutation but are still able to remove the malignant cells.

The main use of beta particles is actually in the field of archeology and geology. The carbon dating technique measures the rate of decay of carbon-14 into nitrogen-14 by emitting beta particles and uses that to determine the age of the fossil, rock, or artefact. This process, including the concept of half lives, will be explored in more detail in chapter 3.

Similar to the rest, gamma rays are also used in the treatment of cancer. However, this can have some rather damning consequences on the rest of this body when a person undergoes chemotherapy. Gamma rays are used frequently in sterilisation. Medical equipment and food are treated with these rays to kill any microorganisms.

We see that this field of radioactivity has many pros and cons. We know the great benefits of using this source of energy, but we must be cautious and act in moderation.

This section has mainly talked about emissions from the nucleus, involving the protons and neutrons, but what about outside the nucleus? What source of energy can we obtain from the electrons? This is where the field of spectroscopy enters our story, with all its quantum effects, spectral lines, and even its role in photosynthesis and respiration.

The Highs and Lows of Energy

Let us imagine a green plant. We know that in order for the plant to survive, it has to produce its own food (glucose), which it can then turn into energy (ATP). Glucose is produced in photosynthesis and ATP is produced in cellular respiration.

Some of you may have heard the phrase 'there is no free lunch'. It essentially demonstrates a total conservation of energy and matter. These two quantities are closely connected. Energy that is lost from matter, is radiated in the form of photons (pure energy) which transfers this energy between matter. We cannot obtain something, without giving up, or using, something else. We cannot create, destroy, or lose any of these quantities in a reaction. The starting and ending amounts must be equal.

This begs the question, where do plants obtain the energy to produce their own food, and how does it happen?

Now, the entire workings of the photosynthetic and cellular respiratory process is far more detailed than needs be included in this book. For our use, we will only focus on the relevant parts.

Firstly, plants obtain energy from the Sun. What does this mean? If plants can obtain energy from the Sun why can't animals do so as well? Green plants have specialised pigments called chlorophyll which traps sunlight in the chloroplast. This light, which is essentially photons, is then used to excite electrons to higher energy levels. There it is. The entire focus of this section. We just learnt about the energy obtained from the nucleus of an atom, but what does it mean to 'excite an electron'?

Once these electrons are excited to higher energy levels, by absorbing photons, they enter the electron transport chain. Think of this transport chain like a hydroelectric dam. Water from a higher area flows downwards which in turn produces energy by rotating turbines, etc. Similarly, the electrons gradually fall down the energy levels, passing from molecule to molecule. During each fall, or step, the electron loses or releases energy which is then used to produce ATP. This ATP is used in a light independent cycle, to produce glucose.

Once glucose is obtained, the same cycle happens again. Only this time, the electrons are not excited by sunlight. The electrons in glucose are already at higher energy levels due to the bonds between atomic nuclei. This is not important to our study, we want to study the interaction between light, electrons, and energy levels.

Thus, we can track the complete energy chain from start to finish. Nuclear fusion in our Sun produces energy in the form of gamma rays. After 100,000 years inside the Sun, these gamma rays have lost enough energy to become UV, visible light, or infrared. These photons are then radiated and hit the Earth. Sunlight is trapped by chlorophyll and the photons excite electrons to higher energy levels. The electrons pass through different intermediary molecules as they fall down the chain and release energy. The released energy is used to produce glucose which is then further oxidised to replenish the energy previously used, and for other living processes. This is the complete energy, electron, and light interaction cycle in plants. But how do we know all this? And how can this same principle be applied for us to study the composition of far away suns and planets?

We will tackle the former question first. This entire section is based on the interaction between two things; electrons and photons. In an atom, electrons can occupy a number of 'energy levels'. There is the ground state, where the electron is closest to the nucleus, and it has the least energy. And there are the excited states where the electron has more energy and is further away from the nucleus. For an electron to jump to a higher energy level, it has to absorb a photon corresponding to that exact energy difference between the levels. For example, an electron at ground state in a hydrogen atom has an energy of -13.6 eV (electron volts). The first excited level has an energy of -3.4 eV. For this electron to jump from ground state to the first excited level, it needs to absorb a photon with energy of 10.2 eV exactly. No more, no less. This is how the electrons in photosynthesis are 'excited'.

Of course, if an electron absorbs energy and becomes excited, it also becomes unstable. The natural tendency for electrons is to remain at ground level, or the lowest energy state possible if there are already too many electrons occupying ground level. To do this, it emits the absorbed energy, in the form of a photon, and falls back to a lower energy level.

The measurement of these energy levels using colour, is called spectroscopy.

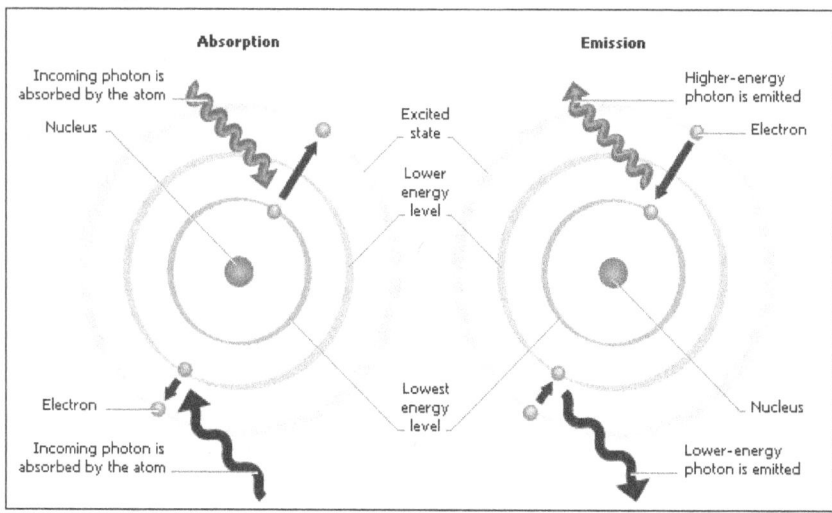

The first evidence of light consisting of 'colour' was the famous prism experiment done by Isaac Newton. Newton split up white light using a prism and observed a range of colours spanning from red to violet, which could be recombined to produce white light again. The colours observed by Newton are the same colours of the rainbow that we see today. Each colour represents a section of visible light with a different frequency and wavelength.

In the early 19th century, this work was improved on by William Hyde Wollaston and Joseph von Fraunhofer. Wollaston built a spectrometer which focused the Sun's light on a screen. When he observed the spectrum of colours, he noticed dark bands of 'missing segments' or gaps in the spectrum. Wollaston believed that these dark lines were the borders between the colours but his hypothesis was ruled out just a decade later by Fraunhofer.

Fraunhofer further improved the experiment by replacing the prism with a diffraction grating. This grating allowed him to improve the spectral resolution, and allow for these dispersed wavelengths to be quantified and given values. By doing so, it was observed that many of these spectral lines, also called Fraunhofer lines, matched the spectra obtained from different laboratories worldwide using different instruments. Clearly, there was some common relationship between these dark bands in the spectrum. But what did this mean?

I mentioned earlier that when an electron jumps down to a lower energy level, it emits a photon corresponding to the difference in the energy levels. So using that stipulation, it can be understood that if an electron is at a higher energy level, it will emit a more energetic photon when it drops to a lower energy level. For example, an electron jumping from the 6th energy level to the ground level emits a much more energetic photon than an electron that jumps from the 3rd level to ground level.

This energy of the photon can correspond to the frequency of visible light emitted. Remember that the energy of a photon is perpendicular to its frequency. If the photon is more energetic, it will have a higher frequency. For visible light, the frequency increases from red to violet. This means that an electron which travels from the 6th energy level to the ground state will emit a photon of the violet band, and an electron that travels from the 3rd to ground state emits a photon in the orange or red band. The same is true for the absorption of photons. A higher frequency, more energetic photon, with a colour closer to violet, will cause the electron to jump to a higher energy level.

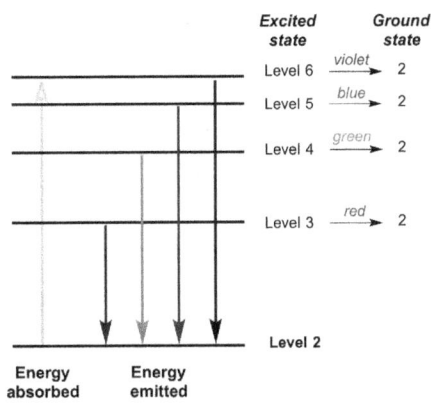

Now, the similarity of spectral lines observed by Fraunhofer is due to the different energies between electron levels that each element has. For example, hydrogen has a certain energy difference between ground state and 2nd level which corresponds to that photon and wavelength in the spectrum. Helium will have a different energy difference between ground level and the 2nd level which will correspond to a different wavelength and energy of photons, and hence a different position in the colour spectrum. Using this, once the natural spectral emission and absorption of each element was determined in the lab, it was easy to determine the constituent elements in a substance just by comparing their spectral lines to the known spectral lines of elements.

The actual notion of energy levels and ground states was only proposed in 1913 by Niels Bohr. but the use of spectral lines to identify elements had been used for decades before that. Bohr just explained how it all worked and what it meant. The quantum mechanical theory of the energy levels and the interaction between the electron and photons was explained in 1926 by Werner Heisenberg and Erwin Schrödinger.

There are two types of spectral lines; absorption and emission. Let us imagine some light source radiating photons. This light source contains the full spectrum of colours, from red to violet. If those photons come into contact with an object, some of the photons will be absorbed by the electrons in the object, causing the black 'absorption lines' in the spectrum. The wavelength and hence colour of the photon absorbed depends on the type of element

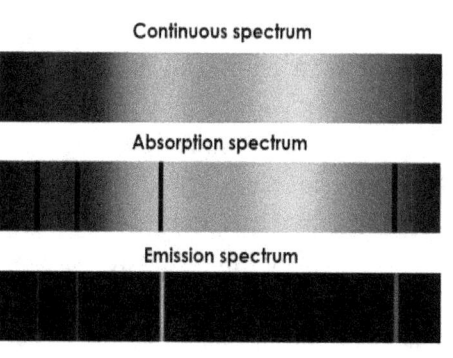

it has come into contact with. When these electrons absorb the photons, they become energetic and unstable. They quickly emit the photons and fall back to a stable energy level or ground level. The emitted photons are called the emission spectrum. This appears as a narrow band of colour on a black background. If we combine the absorption and emission spectra, we get the continuous spectrum of colours that the original photon had. By observing the wavelength of the dark absorption lines and colour of the emission lines, we are able to identify the exact element which makes up the object.

So we now have a method of identifying elements anywhere in the universe, just by observing their spectral lines. The spectral lines of an element is known as the 'fingerprint' of the element. In fact, the element helium was first observed from the Sun's spectrum, giving it its name helium, after the Greek god of the Sun, Helios.

This is how astronomers are able to determine the constituent elements in a cloud of interstellar gas, or an exoplanet's atmosphere. They observe the spectral lines of the gas or atmosphere which has absorbed and emitted the light from its star, and compare it to known values and wavelengths in the laboratory. This is also how the concepts of redshift and blueshift are studied, as light from distant galaxies is either stretched or compressed, so the wavelength of the spectral lines changes. By determining how much it has changed, scientists can work out how far away the source is, and how fast it is travelling, either towards or away from us.

We have learnt all of this, simply by studying the relationship between the electron and the photon. Ideas of energy levels, spectral lines, emission and absorption. How these energetic electrons give rise to vital processes such as photosynthesis and cellular respiration by providing energy. How we can determine the elements present in a star or planet billions of light years away simply by observing its spectrum. This is the power of electrons and light.

So ends our study on the chemistry of the universe. We initially studied the elementary particles that make up all matter. In this, we learnt that, fundamentally, we are all made of the exact same stuff, whether living or nonliving. We then learnt of the four fundamental forces that hold the universe and everything in it together. We continued our story of the Big Bang with the production, annihilation, and asymmetry of matter and antimatter.

Exploring the Essence of Everything

The next three sections focused on nucleosynthesis; how the primordial nuclei were formed, how this process continued in the cores of stars, and how it ended in the explosive death of stars. Artificially, nucleosynthesis continues today as elements are produced in laboratories.

Finally, we explored the power of the atoms. From radioactive decay, fusion reactions, and fission reactions in the atoms, we learnt of the vast consequences, both good and bad, from this nucleic activity. And we ended with a word on electrons and light, as we understood the importance of spectral lines, energy levels, and its quantum consequences.

Journeying on from the microscopic chemical world, we venture into the macroscopic physical world. Now we focus on our own special place in the universe. The only home we have ever known, but hopefully not the only home we will ever know. How can we break down 4.5 billion years of planetary evolution? How can we portray the millions of life forms that have walked, swam, and flown on Earth in one single story? How can we determine the age of rocks and fossils on the Earth? Why was our planet chosen amongst all the rest as a safe harbour for life, that too intelligent life? We have explored the essence of our universe, we have explored the essence of our atoms. For the last chapter in this volume, let us explore the essence of our Earth.

Sections of the Large Hadron Collider at CERN which is used to accelerate particles to near light speed. The LHC has not only discovered the Higgs boson particle, but is also trying to solve a number of unsolved mysteries relating to the beginning and origins of the universe. (CERN)

Murray Gell-Mann (top) and George Zweig (bottom) proposed the quark model in 1964, which was later discovered and proved in 1968 at the Stanford Linear Accelerator. Gell-Mann and Zweig initially worked independently on this theory, with Gell-Mann introducing a symmetry scheme in 1961, requiring the existence of three new elementary particles that he called 'quarks'. Zweig, during a visit to CERN in 1964, added to this theory by saying that mesons and hadrons are constructed from quarks.
Top - (University of Chicago). Bottom - (California Institute of Technology)

J. J. Thomson was a British physicist, credited with the discovery of the electron, the first elementary particle to be found. Thomson made this discovery in 1897 after experiments on cathode rays, owing their magnetic deflection and electrical charge to the electron. (The Royal Society)

Satyendra Nath Bose was an Indian theoretical physicist, who developed the theory of bosons with his colleague, Albert Einstein. The boson is named after him, in recognition of his work in the quantum field. (London, 1925)

Max Planck (top) and Arthur Compton (bottom) both made vital contributions to the development and observation of the photon. In 1900, Planck published his black-body radiation hypothesis, stating that systems absorb and radiate 'quanta' of energy. This is usually known as the birth of quantum mechanics. In 1922, Arthur Compton became the first person to observe the momentum of a photon, a discovery known as the "Compton effect". Both these physicists received the Nobel Prize, Planck in 1918 and Compton in 1927.
Top - (Hugo Erfurth)
Bottom - (Washington University Libraries)

Carlo Rubbia (top) and Simon van der Meer (bottom) shared the Nobel Prize in 1984 for their discovery of the W and Z bosons.
They conducted experiments in 1983 using the Super Proton Synchrotron at CERN. The bosons were produced and observed during high energy proton-antiproton collisions.
Top and bottom - (CERN)

Hideki Yukawa was a Japanese theoretical physicist who predicted the existence and mass of the pion, the carrier of the strong nuclear force that holds atomic nuclei together. When this particle was observed in 1947, Yukawa became the first Japanese Nobel Laureate for his work on the pion and strong nuclear force. (Yousuf Karsh)

Enrico Fermi proposed the first theory of the weak force in 1933, then known as Fermi's interaction. He described the process of beta-decay, one of the three forms of radioactivity. (American Institute of Physics)

James Clerk Maxwell was a Scottish physicist, responsible for the classical theory of electromagnetism, one of the four fundamental forces. Maxwell's theory, published in 1865, unified electricity, magnetism, and light as different manifestations of the same phenomenon. Maxwell's four partial differential equations form the core of electromagnetism, optics, and circuits. (American Philosophical Society)

Isaac Newton (top) and Albert Einstein (bottom) had profound impacts in the study of gravity. Newton first published the basic theories and equations regarding gravity in 1687, in his groundbreaking book *Principia Mathematica*. 220 years later, a young assistant at the Swiss Patent Office in Bern, developed the theory of general relativity. Einstein officially published his theory in 1915, stating that the fabric of spacetime is curved by the matter and energy in it.
Top - (Godfrey Kneller)
Bottom - (Ferdinand Schmutzer / Adam Cuerden)

Paul Dirac (left) and Carl Anderson (bottom) made extensive discoveries in the field of antimatter. Dirac kick started the work on antimatter by publishing a paper in 1928. In it, he formulated the existence of positrons (anti electrons) based on Schrödinger's wave equation. 4 years later, Anderson successfully carried out experimentation to find these antimatter particles using cosmic rays. In light of these achievements, both physicists shared the Nobel Prize in 1936. Left - (Florida State University) Bottom - (California Institute of Technology)

Ralph Alpher was the first person to present the idea of Big Bang nucleosynthesis. He argued that hydrogen and helium were created during this time, explaining their abundance in the universe compared to all other nuclei.
(Alpher Papers)

George Gamow, a Soviet and American physicist, derived the 'Gamow factor', used to determine the probability that two nuclei will undergo nuclear fusion.
(George Washington University)

Hans Bethe analysed the different ways in which hydrogen could turn into helium, and proposed two different methods of nuclear fusion in stars. Bethe was also the head of the Theoretical Division at Los Alamos during the Manhattan Project. (Los Alamos National Laboratory)

Arthur Eddington was an English astrophysicist who first brought light to the idea that stars obtain their energy from nuclear fusion. He published his theories in a 1920 paper and was proven correct a couple decades later. Eddington has a lunar crater and asteroid named after him. (George Grantham Bain Collection)

Fred Hoyle (top) brought the story of stellar nucleosynthesis to its close by detailing how the elements from carbon to iron are synthesised in stars. A few years after his theory was published, Hoyle met a group of theoretical physicists at Cambridge; Margaret Burbidge, Geoffrey Burbidge, and William Fowler (shown in the image from left to right, with Hoyle at the end). This quartet produced the B^2FH paper in 1957 and organised the complete field of nucleosynthesis.

Top - (Institute of Astronomy, Cambridge)
Bottom - (CERN)

Wilhelm Röntgen was a German physicist who produced and detected the first X-rays in 1895. Röntgen won the 1901 Physics Nobel Prize for this discovery. He also has a minor planet and element named after him.
(University of Munich)

Henri Becquerel was a French physicist who was known for discovering radioactivity after experimenting with cathode rays and phosphorescence. Becquerel shared the 1903 Nobel Prize with the Curies for this research.
(Paul Nadar/Adam Cuerden)

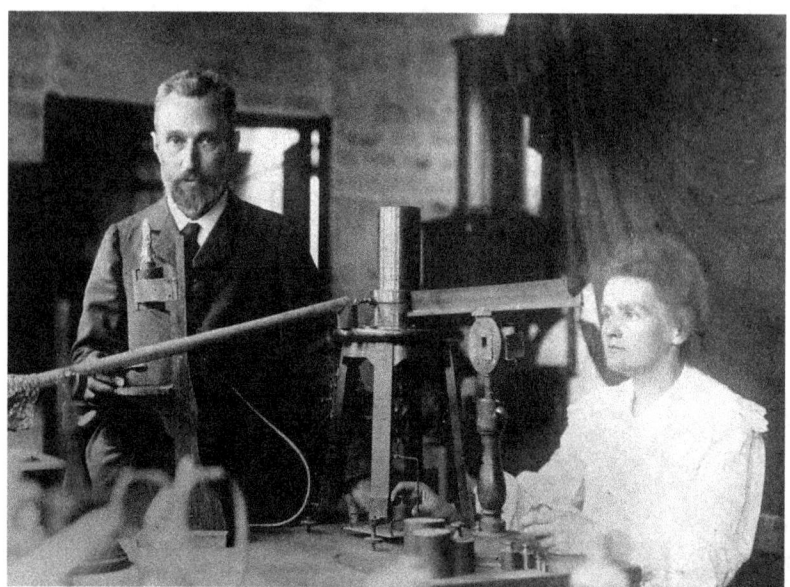

Pierre Curie (left) and Marie Curie (right) conducted experiments on uranium minerals and detected the presence of charged particles in the air, emitted by the sample. The Curies discovered two more elements and shared the Nobel Prize in 1903 with Henri Becquerel. Marie Curie went on to win a second Nobel Prize, this time in Chemistry.
(ESPCI Paris)

Ernest Rutherford (right) was a physicist of New Zealand origin and was the first person to discover the proton after bombarding samples of gold foil with alpha particles.
(Bain News Service)

Joseph von Fraunhofer was a physicist and optical lens manufacturer who also invented the spectroscope. Fraunhofer is famous for his discovery of dark absorption lines in the Sun's spectrum, now known as 'Fraunhofer lines'.
(University of Erlangen)

Niels Bohr was a Danish physicist who postulated the Bohr model of the atom in 1913. This is one of the models used today to describe the movement of the electron in an atom including all its quantum properties. Bohr received the Nobel Prize in 1923 for his work on the quantum world.
(American Institute of Physics)

Erwin Schrödinger (top) and Werner Heisenberg (bottom) made considerable improvements to the quantum model. They used the idea of probability and electron clouds to predict the property and behaviour of the electron. While both scientists had different interpretations and equations for the quantum states of a particle, by the end of the 1920s, they had invented the new quantum theory of Physics.
Top - (Nobel Foundation)
Bottom - (Bundesarchiv, Bild)

3

History of Earth

*"In this broad Earth of ours,
amid the measureless grossness and the slag,
enclosed and safe within its central heart,
nestles the seed perfection"*

Walt Whitman

Exploring the Essence of Everything

Aetas Terrae

Since the majority of this chapter deals with years, eras, eons, and time, it is only fitting that we should return to our Cosmic Calendar. Recall that the universe is 13.8 billion years old, and the Earth just 4.5 billion years old. In terms of the Earth, the first 8 months of the Cosmic Calendar are completely inconsequential, apart from the continuous birth and death cycle of stars, enriching the cosmos for the eventual formation of the solar system. The Earth forms on September 2nd, alongside the rest of the solar system, after a lengthy period of accretion.

Alongside the Cosmic Calendar, we will use another (and perhaps more relatable) timescale to visualise the history of Earth. This is known as the geologic time scale, and it primarily uses the rock record, age of fossils, study of rock layers, and strata to determine the processes that shaped the Earth. The exact methods on how these rocks and fossils are used to determine the age of things shall be explained later, for now we will explore the timeline of the Earth, from a hellish inferno to its current utopia (perhaps an exaggeration).

First, a brief word on the development of this time scale. Where geology and history is concerned, most people think of Harrison Ford, femme fatales, golden idols, and swashbuckling adventures. While the cultural importance of these films is not to be understated, and it has indeed impacted my own life, the science of archeology concerns the history of human relics and artefacts rather than the Earth itself. For that, think *Jurassic Park*. The study of life processes on an ancient Earth using mere moulds and casts in the soil. This is truly groundbreaking science.

This measurement of time and history using rocks stretches long into the BC with observations from the philosophers of ancient Greece. An initial theory was made by Xenophanes of Colophon based on the relationship between fossils and the moving sea bed. Aristotle backed up these claims by reasoning that the position of land and sea must have changed over a long period of time.

A thousand years later, this work was expanded by the Chinese naturalist Shen Kuo, and the Persian polymath Ibn Sina. However, stratification and the recognition of fossils as 'petrifications of the bodies of plants and animals' had little effect on the minds of scholars in ancient Europe, who referred to the Bible to explain the origins of ancient life and fossils. This was the European view until the stipulations of Leonardo da Vinci who strengthened the relationship between stratification, changing sea levels, and time.

The use of rocks to actually determine the age of the Earth and its many changes, was pioneered by James Hutton and Arthur Holmes, born 166 years apart. James Hutton, given the moniker 'Father of Modern Geology', was intrigued by the rock formations he saw. He proposed the geological term 'uniformitarianism', the concept that the present is the key to the past. Hutton published his theories in many books, the most famous of which, *Theory of the Earth,* laid the foundations of geology, showing that everything ancient about the Earth could be learned by studying the patterns, shapes, layers, and components of the rocks.

The notion of uniformitarianism was further popularised by the Scottish geologist Charles Lyell. In terms of geology, Charles Lyell had his finger in every pie, so to speak. Lyell, operating in the mid 19th century, provided influential contributions on Earthquakes, climate change, and the volcanoes Vesuvius and Etna. Lyell also penned 5 separate books across a range of subjects from geology to evolution to an account of his own personal letters and journals. In fact, Charles Lyell was a close friend of Charles Darwin, and contributed significantly to Darwin's magnum opus *On the Origin of Species.*

While James Hutton worked in the realm of theory and idea, Arthur Holmes put in the hard yards and carried out actual physical tests of rock samples to determine their age. This science is called geochronology, using the natural signatures in the rocks and fossils themselves to determine their age, how they formed, and how they changed over time. Holmes's first estimates of rock age were around the 1.6 billion year mark, but did not speculate about the Earth's age. Over the next few decades, he refined his experiments and consequently the results. In 1927, he achieved an age of 3 billion years, and in 1940 he struck gold with a figure of 4.5 billion years, give or take 100 million years.

So where does this place us? Crucial to our study of a time scale, Arthur Holmes developed the modern geological time scale in 1911. This time scale is split into a few sections, each of which are divided into more sections and so on. There are five main divisions in the geologic time scale; eons, eras, periods, epochs, and ages.

Eons are separated into four categories; Hadean, Archean, Proterozoic, and Phanerozoic. The Hadean eon entails the first 500 million years of the Earth's existence, the Archean eon spans the next 1.5 billion years, the Proterozoic eon is the longest, covering a timeline of almost 2 billion years, and our current eon, the Phanerozoic, covers the last 500 million years of Earth's history. We will go through each eon in detail throughout the course of the chapter, studying changes in plate tectonics, the atmosphere, continent formation, and life development.

Eras are split into ten sections, but we will only discuss the last three; Paleozoic, Mesozoic, and Cenozoic. These three eras are concerned with the Phanerozoic eon, the current and most documented length of time in Earth's history. As a result, the majority of our study will concern the Phanerozoic eon, and its three eras.

Moving on to periods, we will explore the latter stages of the periods on Earth, or the most recent in our history. There are 22 periods altogether, of which, we will discuss the last 12, all in the Phanerozoic eon. Worry not, a diagram of the complete geologic time scale is provided at the end of this section.

Due to the detail and length of epochs and ages, they will not be discussed in this book. However, additional reading material can be found in the references. In total, there are 37 defined epochs, and 96 formal ages in Earth's history.

With that, let us begin our dissection into the history of our beloved planet. We start at the gates of hell, in a completely unfamiliar and devastated world, where surely no life can survive, or ever evolve.

History of Earth

Exploring the Essence of Everything

Geologic Time Scale

EON	ERA	PERIOD		EPOCH	
Phanerozoic	Cenozoic	Quaternary		Holocene	Present
					0.01
				Pleistocene	
					2.6
		Tertiary	Neogene	Pliocene	
					5.3
				Miocene	
					23.0
			Paleogene	Oligocene	
					33.9
				Eocene	
					55.8
				Paleocene	
					65.5
	Mesozoic	Cretaceous			
					145.5
		Jurassic			
					199.6
		Triassic			
					251
	Paleozoic	Permian			
					299
		Carboniferous	Pennsylvanian		
					318
			Mississippian		
					359.2
		Devonian			
					416
		Silurian			
					443.7
		Ordovician			
					488.3
		Cambrian			
					542
Precambrian	Proterozoic				
					2500
	Archean				
					4000
	Hadean				
					4500

History of Earth

Hell on Earth

Without delving into too much religious dogma, the concept of hell has been, and will be, prevalent throughout human history. In his 14th century epic poem, *The Divine Comedy,* Dante Alighieri describes hell as nine concentric circles of torment located within the Earth. It is believed that his poem acts as an allegory, representing the journey of the soul towards God, with *Inferno* (the first part of the poem) describing the recognition and rejection of sin. John Milton's *Paradise Lost* portrays a story in hell concerning fallen angels, human temptation towards sin, and a war over Heaven. The underworld has even been mentioned as far back as the BC, with tales of Tartarus and Hades in Virgil's epic, the *Aeneid*.

Spiritual beliefs aside, what can be universally agreed upon is the physical characteristics of hell. It is rather hot. This stems from the belief that since hell is located below the surface, any distance closer to the core will result in higher temperatures. This has resulted in depictions and visualisations of a land of dust, lakes of magma, an infinite din of thunder and bombardment, and a never ending fiery brutality, where nothing can survive. This is known. What if I said that the surface of Earth was also once victim to these hellish circumstances? This is also known.

Dubbed the 'Hadean Eon', the first stage in the geologic time scale of Earth takes its name from Hades, Greek god of the underworld. The term was coined by American geologist Preston Cloud in 1972, during his study of the earliest rocks on Earth.

Let us start with a familiar topic; accretion. Remember that accretion is what allowed the stars and planets to form solid, spherical objects from the dense nebula. With the help of gravity and centrifugal forces, a spherical Earth soon took shape. This primordial Earth would have been extremely hot, with flowing rivers of lava and a molten surface due to the high kinetic energy, friction, and proximity of the recently accreted particles.

Another contributor to the high temperatures and clouds of dust were constant asteroid impacts. Due to the processes of planet formation and accretion, there was a lot of leftover material. In the current day, most of this debris is found in the asteroid belt between Mars and Jupiter and the Kuiper belt past Neptune. So often did these asteroids smash into Earth, and with such great velocities, that one of them even caused the birth of a smaller celestial body. The object we see in its full form every 28 days, which is responsible for our tides and stable climate. The Moon was born from the Earth.

The tale goes that there was once a Mars-sized planet named Theia. Theia travelled through the solar system 4.5 billion years ago during this period of mass formation, orbiting the Sun fairly close to the Earth. During one fateful orbit, the paths of the Earth and Theia coincided, causing a head-on collision and ejecting debris into the nearby vicinity. It isn't said that Theia was completely destroyed or that the Earth lost most of its volume. Rather, the current hypothesis is that the Earth now consists of a small percentage of Theia, while the debris that was ejected, (a mixture of the Earth and Theia) eventually formed the Moon. This explains why the Earth's core is larger than expected for a planet of its size.

This also gives significance to Theia's name. In Greek mythology, Theia was one of the titans, who married her brother Hyperion. They produced a daughter, Selene, the goddess of the Moon. It can be said that Theia is the mother of the Moon, destroying herself in the process to create her offspring.

Speaking of the Earth's core and accretion, there was another natural phenomenon at work during this period. We know that a brick sinks in the sea, while a leaf floats on its surface. Thus, we are familiar with the simple concept of density. Similarly, the Earth at this time consisted of a mixture of heavy, dense, radioactive elements such as plutonium, uranium, and thorium, heavy and dense metals like iron and nickel, and a concoction of lighter elements but in higher abundance.

Using the concept of density, the denser materials such as iron, nickel, and the radioisotopes, would have moved inwards to the centre of the Earth. The lighter elements stayed closer to the surface. However, not every single bit of iron migrated to the core. If that was the case, we would all have severe iron deficiencies and a hazardous lack of oxygen in our bodies. The movement of heavy elements to the core is one of the reasons why life mainly consists of lighter elements such as carbon, hydrogen, nitrogen, and oxygen.

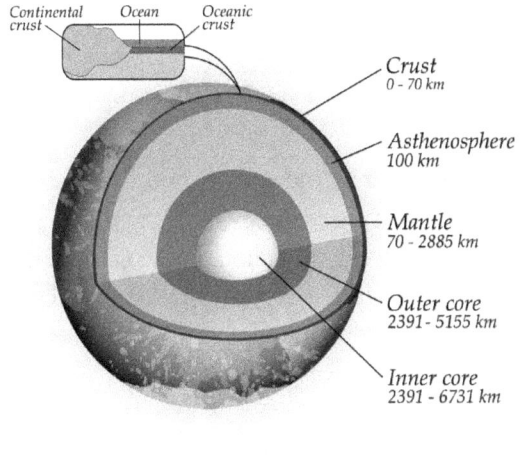

In total, the Earth has 4 layers. Right in the centre, the solid, and dense, hot, is the inner core. The inner core is roughly 85% iron, 14% nickel and 1% other elements. This is surrounded by a liquid layer called the outer core which is kept from solidifying due to the heat emitted by the inner core. The majority of the Earth's mass (68%) is found in the third layer, the mantle. The mantle is made from solid rock with iron and magnesium rich silicates. Between 100 to 350 km underground, the mantle forms a melted rock layer known as the asthenosphere. Geologists believe that this slippery and hot part of the mantle is what tectonic plates slide and move on. Finally, the uppermost layer is called the crust. Even though the crust is all humans have ever known, and most likely ever will know (sorry Jules Verne), it only consists of 1% of the Earth's total volume.

Think of the heights of Everest and the depths of the Marianas trench. These are just the tiniest fractions of the natural wonders the Earth has revealed to us. What's to say there isn't a whole other civilization living inside the Earth ala *The Matrix* or *The Penultimate Truth*? In actual fact, the heat and pressure below the surface far exceeds the limits of human tolerance. But maybe we can't rule out other evolved life forms. After all, it took us 4.5 billion years just to be able to live and function on the surface.

With these frequent asteroid collisions, there were considerable consequences and effects on a planet just in its infancy. These changes on a newborn Earth shaped the very face of our planet in terms of its land, sea, and air.

When we imagine Venus today, we think of a carbon dioxide rich atmosphere, several hundred degrees hot, with pressures that would crush an African elephant. Funnily enough, the Hadean eon on Earth was not so different. Each asteroid impact on the surface would produce copious amounts of heat and hot volatiles. This resulted in an atmosphere dense with carbon dioxide, water vapour, and hydrogen. Most of the water vapour would have been broken down by UV rays from the Sun, producing oxygen and more hydrogen. The majority of the hydrogen would have left the atmosphere due to its low density, while the oxygen would be removed by the reducing atmosphere. This left an atmosphere full of carbon dioxide, producing a near runaway greenhouse effect, with temperatures of 230°C and a pressure of 27 atmospheres (for comparison, our current atmospheric temperature and pressure is 20°C and 1 atmosphere).

You would think that these appalling conditions would write off any chance of liquid water on the surface. You would be wrong. Most people assume that the physical state of water at any one time is temperature dependent. This isn't wrong, it's just half of the complete picture. We say that water boils at 100 °C, and 1 atmosphere. The pressure is absolutely key. The whole concept of 'boiling' is just the ability of the water particles to gain enough kinetic energy to not only overcome the forces of attraction between each other, but also the pressure of the atmosphere pushing down on them. This is why boiling a liquid takes much more time than melting a solid. Melting is not as pressure dependent. This is also why water on Mars vaporises so easily regardless of the near zero temperatures, the lack of atmospheric pressure makes it easy for the particles to escape into the 'air'.

So even though the early Earth temperature was double the current boiling point of water, causing the particles to move with intense kinetic energy, the atmospheric pressure was 27 times that of normal boiling conditions, creating such a crushing pressure on the particles, preventing them from becoming water vapour. The primordial 'oceans' would have absorbed some carbon dioxide from the atmosphere but not enough to substantially reduce its amount.

Compared to the atmosphere and the oceans of water, research on early Earth continents has so far proved futile. The main consensus today on the existence of the first continents depends on the water level and the amount of continental crust. It is believed that while the continental crust continued to grow over the 500 million years of the Hadean eon, the ocean levels became increasingly deeper due to water being released from the mantle. This could be one of the reasons why life started in the oceans rather than on land.

Overall, we see that the Hadean eon was a time of intense reckoning for the early Earth. A 500 million year span of constant bombardment, producing the Moon alongside tons of heat and carbon dioxide. Eventually, the elements sorted themselves out as the layers formed, sea levels rose as did the continental crust, and the benefits of an intense atmospheric pressure prevented the loss of all liquid water from Earth's surface. Following these troubling times, the Earth began to emerge from its traumatic childhood into a more calm and conducive adolescence.

Exploring the Essence of Everything

The Calm After the Storm

In Tom Cruise's 2018 blockbuster, *Mission Impossible: Fallout*, there is a piece of dialogue from the villain, Solomon Lane, that often appears in my thoughts. When he is doing the usual 'bad guy explains devious plan and motive' monologue, he says the phrase *"There cannot be peace without first, a great suffering. The greater the suffering, the greater the peace."*. Of course, when it is said in the movie, it's in the context of using nuclear warheads to wipe out a third of the world's population. However, the concept of purging to purify and 'restart' with better ideals has been hinted at before, and is indeed prevalent throughout history.

We are all familiar with the story of Noah and his ark. How an omnipotent being rains down a flood from the heavens to purge the world from the sin and corruption of man. Noah, a pious and pure individual is instructed to build an ark to save his family and two organisms (a male and a female) from every animal species on Earth. After 370 odd days aboard the ark, Noah, his family, and all the living creatures disembark at the mountains of Ararat, (in present day Turkey/Armenia), preparing themselves for a reborn and better world. I suppose whether you believe in the truth of the story is besides the point, but once again, there is a lesson there to be learnt.

The idea of a great calm after a great storm is shown in the very foundations of the Earth. However, the handing of the torch from the Hadean to the Archean eon was not a straightforward one. Rather, it was a more drawn out process, culminating in one final deluge of torment from the heavens.

When the Late Heavy Bombardment (LHB) did arrive, it did not discriminate. The asteroids rained down on all the terrestrial planets in the solar system, not just the Earth. Even though this event occurred around 4 billion years ago, it wasn't until the last 50 years that the idea of it gained traction. This was mainly due to human exploration of our closest celestial neighbour.

History of Earth

Between 1969 to 1972, six Apollo missions brought back 382 kilograms of lunar rock, core samples, sand, and dust, across six different lunar sites on the surface. Among these rock samples were impact melt rocks. By studying the age and formation of these rocks, scientists were able to deduce that there must have been a period of heavy asteroid bombardment and impact collision with the lunar surface, and the subsequent terrestrial planets, during the passing of the Hadean eon into the Archean eon. Thus, the Late Heavy Bombardment is often known as the lunar cataclysm.

A number of potential causes for the LHB have been cited. Why did the debris of space suddenly spike in magnitude and quantity after its initial intense period of bombardment? Some scientists have accredited this to the migration of Jupiter towards the inner solar system, disrupting the motion of asteroids and causing them to start their willy-nilly orbits once again.

A few other reasons include the late formation of Uranus and Neptune, the breakup of a large main asteroid belt, and the existence of a hypothesised planet V. Planet V could have orbited between Mars and the asteroid belt, becoming unstable after perturbations from the inner terrestrial planets. This would have caused it to careen into the bulk of the asteroid belt, causing many asteroids to enter near-Earth orbits. The existence of planet V is said to have ended in a calamitous fireball, as it plunged into the Sun.

After this period of heavy bombardment came the calm and lull of a devastated Earth. The high number of impact craters gave rise to an increase in volcanism, a process that began in the Hadean eon. The eruption of molten rock from inside the Earth to the surface. As a result, this increased the amount of carbon dioxide in the atmosphere, and further reduced the amount of oxygen. Some analyses have shown that Archean eon oxygen levels could have been as low as 0.0001% of modern atmospheric levels. Whatever tiny fraction of oxygen produced would have been by cyanobacteria in the oceans, carrying out photosynthesis. Although oxygen levels over the next 1.5 billion years did not improve appreciably, the groundwork had been set in the Earth's oceans for an increase in atmospheric oxygen.

The oceans themselves continued to expand and increase in depth due to outgassing from the mantle. This caused the Earth to essentially become a water world, as the oceans covered the continental crust almost entirely. This also caused the abundant magma from volcanoes to solidify, turning them into igneous rocks such as basalt and granite.

The main occurrence in the Archean eon, aside from the Late Heavy Bombardment and emergence of life, was the formation of the continents. The first larger pieces of continental crust consisted of felsic basaltic rock, with deeper layers of iron. What is left of these initial continents are called cratons. Cratons are currently found in the centre of every continent on Earth, and in the interiors of tectonic plates. Using the foundations of cratons, the crust grew outwards and expanded to form the landmasses of today.

It can be said that not a whole lot happened during the Archean eon, when compared to the other three eons of Earth. This was a true period of recuperation and serenity after half a billion years of onslaught. The 1.5 billion years of the Archean eon passed relatively quickly so to speak, as the Earth began to lay the foundations for the development and evolution of life.

Laying the Groundwork

Let us refresh our timeline along the Cosmic Calendar. Previously, I said that the Earth would have formed on the 2nd of September. Since its formation to the end of the Archean eon, (roughly 2 billion years) just 55 days have passed. This means that at our current juncture, it is the 27th of October. Much more has to be done during the remaining two months to get civilization to the point it is at today. Mainly, we need life to

History of Earth

evolve. For this, we need to lay the foundations for life to evolve; we need landmasses to change and become conducive to land based creatures, we need to alter the gaseous constituents of the atmosphere to allow for a more nurturing environment in which life can flourish, we need sunlight to hit the Earth at exactly the right intensity while blocking harmful ultraviolet rays.

Keep in mind that all these conditions are only specific to life on Earth today. Whether the oxygen content increased, landmasses emerged from the oceans, or the Earth had been frozen in a perpetual winter, life would have evolved anyways. But in order for it to be exactly what it is today, certain things needed to happen. This is the story of those things.

As the transition between the Hadean and Archean eon was marked by the Late Heavy Bombardment, so was the transition between the Archean and Proterozoic eon marked by a similarly noticeable impact on the Earth. Compared to the deafening and evident impacts of asteroids during the LHB, sending plumes of dust and smoke into the atmosphere, the Great Oxidation Event (GOE) of the Proterozoic eon was much more subtle in its appearance, and produced mainly a chemical change rather than a physical change.

The GOE occurred some 2.4 billion years ago and ended 400 million years later. Remember from the previous section that cyanobacteria in the Archean eon had begun to emit oxygen into the atmosphere during photosynthesis. This measly amount of oxygen mattered not in the grand scheme of the Archean atmosphere. However, just as life itself is nothing more than the accumulation of lots of little things, this primordial oxygen eventually accumulated into an amount which could discernibly impact the conditions of a Proterozoic Earth, particularly its microbial life.

In the current day, we know that carbon dioxide levels are rising. This increase will not really affect our circulatory system right now, as the ratios between carbon dioxide and oxygen in the atmosphere are vast, but give it a few hundred million years, and we might be talking of a Great Carbonation Event. Naturally, this would kill off almost all life on Earth that has evolved to breathe and survive in an oxygen rich and carbon dioxide deficient atmosphere. The inverse of this situation was a reality 2.5 billion years ago.

When the levels of oxygen in the atmosphere spiked during the dawn of the Proterozoic eon, most of the life that had evolved during the Archean eon, mainly anaerobic organisms, died out and went extinct. This shows how fickle and fragile life can be. Change the chemical composition of the atmosphere by injecting a few divergent molecules, and a mass extinction occurs. But it is key to remember that not all life was wiped out. The surviving prokaryotes would have evolved into aerobic proteobacteria, leading to the rise of eukaryotes with membrane bound nuclei, and the eventual evolution of multicellular life forms.

Apart from triggering the evolution of modern day life, the GOE also helped to reduce the temperature of a sweltering Earth. The oxygen would have oxidised methane, a strong greenhouse gas, into carbon dioxide, a weaker greenhouse gas, causing planetary cooling. The GOE was also responsible for producing over half of the total minerals found on Earth today. Most of these new minerals were formed as a consequence of elements in the dynamic mantle and crust becoming oxidised and hydrated.

Coinciding with the GOE, in fact, as a result of the GOE and a cooling Earth, occurred the Huronian glaciation. Inferred in 1907 by Canadian geologist Arthur Philemon Coleman, this was a period where at least three separate ice ages occurred. Coleman came to his conclusion on analysis of geological rock formations near Lake Huron, Ontario. A popular perception is that the entire surface of Earth at this time was covered by a layer of ice. However, the glacial sedimentary rock (diamictites) are discontinuous, often alternating with carbons, silicates, and other sedimentary rock, hence casting doubt on the idea of a 'frozen Earth'.

So we see that the initial conditions for the evolution of current life had been fulfilled early on in the age of the Proterozoic eon. The GOE occurred just as the Archean eon reached its culmination, resulting in higher oxygen levels for life to evolve into present day beings, and reducing the temperature of Earth drastically. The final piece of the puzzle before life could fully emerge from its cocoon in an adolescent Earth was for the landmasses to emerge from the depths of the ocean.

Out of the four eons, the tectonic activity during the Proterozoic eon was the highest and most active. For example, 43% of the modern continental crust was formed in this period, with 39% being formed in the Archean (the cratons), and a mere 18% during the Phanerozoic. The constantly colliding and shifting tectonic plates caused the landmasses to emerge from the oceans after millions of years of crustal recycling.

This continental drift was first speculated as far back as the 16th century! The Dutch geographer and cartographer Abraham Ortelius, in the year 1596, underlined the similarity between the shape of the coasts of America and Europe/Africa. He suggested that the Americas were torn away from the European and African landmass by earthquakes and floods. He wrote in his *Thesaurus Graphicus* that *"The vestiges of the rupture reveal themselves, if someone brings forward a map of the world and considers carefully the coasts of the three continents."*

Unfortunately, the idea of continental drift was relatively unexplored until 300 years later, when a handful of geologists made more speculations of the viability of supercontinents on a changing Earth. The theory was not fully complete until the work of Alfred Wegener, who formulated his entire continental drift theory in a 1912 paper. The German presented his hypothesis to the German Geological Society on the 6th of January, 1912, stating that the modern day continents once formed a single landmass called 'Pangaea'. He was also the first to coin the term 'continental drift'. However, Wegener was uncertain as to what actually caused the continents to move on their own accord, and put it down to a centrifugal pseudo force from the Earth's movement. Of course, in the present day we know that continental drift, earthquakes, and tsunamis are all caused by tectonic plate movements.

As a result of these tectonic movements, the first supercontinents began to emerge. In truth, a couple of them had already been formed as far back as 3.5 billion years ago. However, these supercontinents would have been mostly submerged in the oceans. The main supercontinents of the Proterozoic eon were Columbia and Rodinia.

The supercontinent Columbia would have formed as several small cratons came together and fused to form one single large landmass. Columbia contained fragments of current day India, Australia, northern China, and the core of North America. The lifespan of Columbia is estimated to have been around 600 million years. Clearly, these tectonic activities take their time in producing noticeable effects, but they do have the power to change the structure of the Earth's surface at will.

As Columbia began to break apart, due to continental drifting, the fragments once again began to accrete and collide to form the supercontinent Rodinia. Rodinia, from the Russian word *rodina*, meaning 'birthplace' or 'motherland', had a similar lifespan to that of Columbia. Eventually, Rodinia would have crumbled and a new supercontinent would have taken its place. And the cycle goes on and on. But the main point is that viable landmass now exists on the Earth's surface, albeit in a constant loop of accretion and disintegration.

And so all the pieces of the jigsaw have been studied and assembled. Our air is filled with oxygen, the temperatures have cooled, and sufficient land has come forth from the oceans. At last, the stage is set for the grandest reveal of them all. The explosion of life.

Let There Be Life

Our final stop on the odyssey that is historia terrae takes us to the Phanerozoic eon. The most recent and current of the four geological eons, it is no surprise that the Phanerozoic contains the most events and timelines for us to break down. As a result, the entire eon will be divided into three sections, one for each of its eras. This section will focus on the Paleozoic era, encompassing the first 287 million years of the

History of Earth

Phanerozoic. The Paleozoic era can further be divided into six periods; Cambrian, Ordovician, Silurian, Devonian, Carboniferous, and Permian.

By far the most famous feature of the Paleozoic era ('early life'), and perhaps the Phanerozoic eon as a whole, is the Cambrian explosion. Unlike most explosions, in which an enormous emission of energy and heat culminates in the destruction of property and the loss of lives, the Cambrian explosion provided a quite opposite impact. This was dubbed 'the explosion of life' due to a stark increase of complex life and diversification in the fossil record.

Prior to this, most life consisted of basic, individual or multicellular cells, with barely any variety. The most famous pre-Cambrian fossils are the stromatolites. The stromatolites themselves were not living organisms, but are actually layered sedimentary formations created by photosynthetic microorganisms such as the aforementioned cyanobacteria. These microbes produced adhesive compounds that 'cemented' sand and other rocks to form mineral 'microbial mats'. These mats accumulated over time, being built up layer by layer. Refer to the images at the end of the chapter for a more vivid idea.

The formation of stromatolites peaked around 1.25 billion years ago, indicating their major role in the fossil record of the first life forms on Earth. However, the oldest stromatolites can be found in Western Australia, dating back a whopping 3.5 - 3.8 billion years ago! Archean stromatolites are relatively rare as the population of cyanobacteria would have been in its infancy. As the Earth healed and more oxygen accumulated, Proterozoic cyanobacteria became more and more common. As a result, the majority of stromatolite fossils originate from the Proterozoic eon.

Apart from its extensive role in the history of life on Earth, stromatolites are not just a thing of the past. Present day saline and inland formations can be found in Shark Bay and Late Thetis, Australia, as well as locations in Mexico, Brazil, and Chile. For those interested in submerged, freshwater locations, giant formations have been found across Mexico's Yucatan Peninsula and a few locations in Turkey and Canada.

So cyanobacteria and stromatolites dominated much of the pre-Cambrian fossil record. Following the evolutionary boom though, it was a rapid jump from the microscopic world of the microbes to the fully functioning, macroscopic world of arthropods, the dominant life form at this time. Arthropods are an extremely diverse group of organisms, with up to 10 million species in their arsenal. Modern day arthropods include insects, shrimp, spiders, and centipedes. The arthropods of the Cambrian period differed, in that they were mostly submerged and simple versions of today's organisms. The other form of evolved life at this time were the molluscs. Examples of these early forms of life can be found in the mid-section of the book.

The appearance of arthropods in the evolutionary chain of life also had an effect on the microbial life that had preceded it. The 'microbial mats' and stromatolites of the cyanobacteria which were previously abundant, now became scarce as burrowing animals destroyed these structures through bioturbation.

Such a rapid change in the features of life then poses the question; what exactly caused this dramatic shift in the physiology and genealogy of life? Perhaps the most influential factor would have been an increase in oxygen levels. As more atmospheric and hence dissolved oxygen became available, more complex organisms could have formed with the beginnings of a respiratory and circulatory system.

Another reason could be the 'snowball Earth' hypothesis previously discussed in the Proterozoic eon. As the Earth warmed up and the ice melted, more sunlight would have been available for cyanobacteria to carry out photosynthesis, further enriching the environment with oxygen. Also, as glaciers receded and rocks eroded, nutrient-rich sediments may have been deposited into the oceans, paving the way for the diversification of life.

Following the end of the Cambrian period, we move into the Ordovician period. Spanning 41.6 million years, the Ordovician period marked a great transition from water based life to land based life. The colonisation of land was made possible as the southern continents amassed into the supercontinent Gondwana.

Life continued to flourish as the invertebrate molluscs and arthropods dominated the oceans. The first arthropods started to colonise land during this period, helped by changes in anatomy and genealogy. If the Cambrian period can be thought of as a leap from simple organisms into larger complex life, the Ordovician period represents the biodiversification of organisms as morphological disparity was similar to todays. Dubbed the 'Great Ordovician Biodiversification Event' (GOBE), it was not just responsible for the change in fauna life, but also an increase in flora life forms. In fact, the first evidence of land plants is from this time. The GOBE also allowed for the evolution of the first vertebrates; fish. Molluscs and arthropods became more varied as trilobites, a now extinct form of arthropod, became widespread and diverse.

The Silurian period spanned just 24.6 million years, making it the shortest geological period in the Paleozoic era. The most important development in this period was the development of terrestrial life, both flora and fauna. The three smaller continents at this time; Avalonia, Baltica, and Laurentia, drifted together near the equator and eventually formed the second supercontinent of the era; Euramerica.

During this time, three groups of arthropods became fully terrestrial; myriapods (millipedes, centipedes), arachnids (spiders, scorpions), and hexapods (flies). Do note that the organisms in the brackets are just examples of each group, not representatives of life on Earth at that time.

In addition, the first vascular plants emerged on land. Vascular plants contain tissues for transporting water, minerals, and food through the plants, although any form of these fauna at the time would have been much more simplified. Be that as it may, it marked a significant change in the development of plant life from the prominent algae in the oceans.

The Devonian period, the fourth period of the Paleozoic era, is known as the Age of Fishes. Jawed fish reached substantial diversity during this time, as placoderms (armoured fish) began dominating almost every aquatic environment. The first primitive sharks and cartilaginous fishes emerged at this time as the tissues and sizes of aquatic organisms became more diversified and complex.

This period also marked the start of the tetrapod (four-limbed vertebrate) line of organisms. Tetrapodomorpha began diverging from freshwater fish as their more robust and muscular fins gradually evolved into forelimbs and hindlimbs. However, colonisation of these organisms on land and the full specialisations of their limbs were not established until the late Carboniferous period.

In terms of paleogeography, the supercontinent Gondwana continued to dominate the southern hemisphere of the Earth. As Gondwana and the smaller supercontinent Euramerica began to approach and collide, the formation of the famous supercontinent, Pangaea, was starting to come to fruition.

By the Devonian period, life had developed a substantial foothold in the colonisation of land. Primordial forests began to emerge as plants developed leaves, woody stems, and roots, and soil became more conducive to terrestrial arthropods. By the end of this period, the first seed-bearing plants had also evolved. The late Devonian also saw the most rapid plant diversification as green forests began to emerge, leading into the Carboniferous period. With all the talk of vascular plants and terrestrial arthropods, it was actually fungi that dominated the early stages of this terrestrial biodiversification event.

The penultimate period of the Paleozoic era, the Carboniferous period, spanned 60 million years and created much of the carbon that became coal deposits of today, hence the name. This carbon was mainly due to the remains of trees, which had grown to greater heights and numbers, owing to the development of lignin to strengthen the trees. However, terrestrial bacteria and fungi at the time had not sufficiently evolved to break down this lignin and as a result, the trees remained mostly buried in soil.

History of Earth

This period is sometimes called the Age of Amphibians as early amphibians became diversified and started to colonise both terrestrial and aquatic habitats. This period also marked the first appearance of the ancestors of mammals, the synapsids, as well as the ancestors of reptiles and birds, the sauropsids. Due to raised atmospheric levels of oxygen from increased photosynthesis, land arthropods underwent major evolutionary changes, as their physiology altered to resemble the arthropods of today.

New groups of land plants also appeared during this period, some of which were huge trees with trunks 30 metres high and up to 1.5 metres in diameter. The different genus and species of plants grew a hundred fold as new types sprung up every few thousand years, leading to a reduction in atmospheric carbon dioxide levels.

This reduction of CO_2 levels caused climate cooling and glaciation along the south polar region, particularly in Gondwana. This cooling eventually led to the Late Paleozoic Ice Age (LPIA), the longest icehouse period of the Phanerozoic, spanning 112 million years. Average global temperatures during the LPIA were as low as 13°C, with extremely varying temperatures in the tropics (24°C) and the polar regions (-23°C).

This drop in carbon dioxide also led to tetrapods acquiring new terrestrial adaptations such as the dryland-adapted amniotes (tetrapod vertebrates evolved from amphibians). These amniotes evolved the crucial ability to lay eggs in dry environments, as well as scales to prevent water loss and hardened claws for further exploration of the land.

The appearance of oversized trees was not the only bizarre physical occurrence during this time as some terrestrial invertebrates grew up to 2 metres in length! In fact, the largest land invertebrate and flying insect of all time were both prominent during this time. The millipede-like arthropod, *arthropleura*, had sizes up to 2.6 metres long, and the giant dragonfly-like insect, *meganeura*, had a wingspan of up to 75 cm wide.

Finally, we reach the last stage in the Paleozoic era, the Permian period. Spanning just 47 million years, the beginning of this period was marked by the complete formation of the supercontinent Pangaea. For every supercontinent there was a superocean and such was the case here, as Pangaea was surrounded by a single ocean called Panthalassa.

The interior region of Pangaea experienced heavy continental climate with extreme variations of heat (a massive desert in its central region) and cold (the LPIA continued during this time) as well as monsoon periods with seasonal rainfall patterns. This caused further evolutionary branching of terrestrial life.

In terms of plants, conifers and other gymnosperms (plants with seeds enclosed in a protective cover) had the upper hand on ferns and other swamp trees that dispersed spores in a wetter environment. The first ancestors of the Ginkgo tree surfaced during this period as the fauna on Earth started to resemble modern day habitats.

Synapsids (which would later include mammals) continued to thrive during these harsh, dry times as they became the dominant vertebrates. As for future species, the first ancestors of the dinosaurs and pterosaurs evolved and diversified during this time, ready to make their mark in the next era of the world. More anatomical modern animals continued to emerge as the first beetles appeared as well as modern amphibians (lissamphibians).

At last, our journey across the Paleozoic era comes to a close. This has been an eye-opening glimpse into the development and propagation of life on Earth. From the oxygen-producing cyanobacteria and algae of the past 2 billion years, and its fossil evidence in stromatolites, all the way to anatomically modern amphibians and fish, it has been a story of evolution and adaptation to ever changing environments. We have seen the emergence of the first invertebrates, the arthropods, to the first terrestrial life that crawled out of the oceans, to the first vascular trees and plant life.

The physical conditions of the Earth have also changed much during these 289 million years, as supercontinents and tectonic plates worked their magic movements over millions of years to eventually form Pangaea. Oxygen and carbon dioxide levels rose and fell like so many tides in the oceans. And the various ice ages and climate changes during this time, forever shaping the evolutionary track of life on Earth. Finally, the path has been cleared for an adventure 66 million years in the making.

Welcome to Jurassic Park

Let me drop a few facts and figures on the types of species found throughout the history of our world. At the present date, we have officially classified 2.1 million different species of plants and animals, spanning from microorganisms 2 micrometres long to 100 foot trees scratching the heavens. It is estimated that in total, there are currently 8.7 million types of species in our world. This means that we have identified and classified just 24% of all species currently alive on the Earth. However, the number of 8.7 million species is fairly insignificant when compared to all the species that have ever lived on the Earth, throughout its 4.5 billion years.

Of course, it is rather difficult to get an accurate number for something which lived millions of years before humans even existed, but following the tiresome work of palaeontologists, it is said that 99.9% of species that have ever lived on the Earth are now extinct. This is a crude estimation but it is the best figure we have at the present for events occurring over the timespan of billions of years. This would mean that only 0.1% of the total number of species to have ever existed on the Earth, are currently alive right now. With our previous estimate of 8.7 million species, this would give a grand total of 8.7 billion species to ever exist on the Earth. And indeed, most palaeontologists and scientists put this estimate around the 5-10 billion mark. For average's sake, let us put this figure at 7.5 billion.

Exploring the Essence of Everything

Alright, let's keep that number in mind. There have been roughly 7.5 billion different types of species to have ever lived on the Earth. Now let's get slightly more specific and talk about a certain type of organism that lived some 250 to 66 million years ago. The dinosaurs themselves, only consisted of just more than 1,000 species. Comparing these two figures is akin to comparing David and Goliath, and funnily enough, the results are also similar. Just as David beat Goliath in the book of Samuel, so does it happen here when we compare the species of dinosaurs against the total number of species on Earth. Let me elaborate.

If dinosaurs only accounted for 1,000 species of the total 7.5 billion species to have ever lived on Earth, it only accounts for 0.00001% of all species. Then why, when we compare dinosaurs to any other prehistoric species on Earth (in this case, prehistoric refers to any time before the emergence of humans), do dinosaurs always come out on top? In terms of popularity, scientific endeavours, field work, museum exhibits, movies, books, and any other form of modern day interest, dinosaurs triumph over any other prehistoric species. What makes this 0.00001% so special? Let us first see how they appeared, thrived, survived, and eventually, died.

The second era of the Phanerozoic eon takes us to the Mesozoic era ('middle life'). This era spanned 184 million years and is divided into three periods.

The first period, dubbed the Triassic, spanned 50.5 million years and is the shortest of the three Mesozoic periods. Life at the beginning of the Triassic was severely impoverished and on its last legs.

The devastating Permian-Triassic extinction event had wiped out nearly 95% of all life on Earth due to violent volcanic eruptions, expelling copious amounts of sulphur and carbon dioxide into the atmosphere. This caused two major changes in the environment. Firstly, temperatures increased rapidly as greenhouse gas content increased, causing the already dry and sterile land to become even more uninhabitable for life. Secondly, the increased levels of sulphur dioxide caused euxinia, a condition where oxygen content in water is depleted, and sulphuric content is increased. This causes the water to become acidified, creating unsuitable ph levels for aquatic life.

As temperatures began to cool during the mid Triassic, the Earth healed once again, this time from an ecological disaster rather than a physical one. Life began to emerge and flourish as ecosystems recovered and conditions became more favourable.

In terms of Triassic fauna, it was not as dire as one might assume. In fact, the lycophytes, vascular plants that reproduce using spores, actually thrived during this period of environmental instability. In particular, the genus *pleuromeia* dominated almost all early Triassic floral life, growing up to 2m in height. Unfortunately for *pleuromeia,* it started to decrease in relevancy and quantity as conditions for its survival became less favourable and competition for resources increased. I suppose it made the best of a bad situation for as long as it could.

If the Proterozoic era can be called the Age of Amphibians, then the Mesozoic era can certainly be called the Age of Reptiles. For all the amphibian groups present during the Proterozoic era, right up to the end of the Permian, only a handful survived the hindrance of extinction. Among these, the most widespread were the temnospondyl amphibians, most of which died out during the early to mid Triassic anyways. One group which did not die out, and in fact has never died out, were the lissamphibians. These modern day amphibians continued to evolve right throughout the Mesozoic, with the progenitors of the first frogs emerging in the early Triassic.

And now, the magnum opus of the Mesozoic. Its crowning jewel. A species so scarce when compared to others that came before and after it, and yet, is still famous today, 66 million years after its untimely demise.

The first cousins of the dinosaurs, just like every other life form on Earth, evolved in the oceans. While these marine organisms were not part of the dino family tree, they can be thought of as distant cousins, since they are both types of reptiles.

The Ichthyosaurs were highly successful during this period, often acting as the tertiary consumers in highly competitive food chains and webs. These predators soon grew to outstanding sizes during the late Triassic and can be thought of as a shark-dolphin hybrid. See images for reference. Alongside the fearsome Ichthyosaurs, there were also the Nothosaurus, growing up to 3 metres in length, and the Plesiosaurs. The Plesiosaurs are fascinating due to the fact that they managed to survive right up to the end of the Mesozoic, far surpassing the lifespan of their cousin marine species. For all these carnivorous marine reptiles, there was also the first herbivorous marine reptile, the Atopodentatus, feasting mainly on algae from the seabed.

Unfortunately for the popcorn headlines, the most commonly known terrestrial dinosaurs were not really present during the Triassic period. In fact, the dinosaurs present on land were mainly small predators such as the Coelophysis and the Staurikosaurus. Most terrestrial reptiles during this time were more lizard and crocodile-like.

The vital clade of reptiles that diversified at this time were the Avemetatarsalia (meaning bird metatarsals), the ancestors of all birds found today. This clade of reptiles split into two very successful groups; dinosaurs and pterosaurs. The pterosaurs were the first winged vertebrates capable of flight, only lasting till the extinction event at the end of the Cretaceous. The dinosaurs however, continue to thrive in today's world and we perhaps take this fact for granted.

If you've ever heard someone say that 'chickens are the closest relatives to the dinosaurs' or that 'all birds are dinosaurs' there is some truth in it, although it is mostly said to grab attention. The fact is that all present day birds evolved from the same group of reptiles from which terrestrial dinosaurs are a part of. As this group continued to diversify, the terrestrial dinosaur branch consisting of the T-rex, Velociraptor, Stegosaurus, etc, would become extinct in that meteor event. However, the group that branched off to become the birds of today, has managed to stay alive 66 million years after the fact. So you can think of birds as the great-great-great grand cousins of the dinosaurs. They share almost no physical similarities, but genetically, there are more similarities than you might expect. This is also part of the reason why most palaeontologists claim that dinosaurs had feathers.

Moving past the Triassic and into the Jurassic, we encounter a break up of monumental proportions. Pangaea is nearing its end. Made up of Gondwana to the south and Eurasia to the north, this once titan of landmass on Earth begins to separate into its constituent continents. As Pangaea split in half and smaller islands broke away, this allowed for large seas to appear in between the landmasses. With water much closer to the majority of land once again, as opposed to the central arid region of Pangaea, the climate was tropical and much more humid compared to the Triassic.

As the habitats changed and flora, particularly conifers, became more prominent, this allowed for further diversification of fauna. This period was the peak of reptilian life, following its emergence during the Triassic and preceding its extinction at the end of the Cretaceous.

In the oceans, the Ichthyosaurs continued to thrive after suffering an evolutionary bottleneck during the late Triassic. The Plesiosaurs became the last of its kind as its cousins, the Placodonts and the Nothosaurus, had become extinct. These Plesiosaurs began to invade freshwater environments, with some fossils found in the freshwater sediments of China and Australia.

The first avian vertebrates, the pterosaurs, evolved during this time. These winged lords of the sky were thought to be piscivorous, with a diet consisting mainly of fish, although some pterosaurs were strictly insectivores, perhaps owing to shorter heads than their piscivorous cousins, that were unable to reach fish in the oceans.

As for the dinosaurs, they became the dominant vertebrates in terrestrial ecosystems. Dinosaurs had truly ruled the Earth. The first modern day bird ancestors, the theropods, evolved during this time. This was definitively represented by the Archaeopteryx, avian dinosaurs that thrived in what is now southern germany. This prehistoric bird was rather small, only reaching a maximum length of 0.5 metres, and was a cross of dinosaur and modern day avian species, with a bony tail, three fingered claws, jaws with sharp teeth, and feathers.

For fans of the *Jurassic Park* franchise, here are a few dinosaurs that might pique your interest. The first armoured Ankylosaurus and spined Stegosaurus appeared during the middle Jurassic. These herbivores were dwarfed (literally!) by the dominant herbivores at the time, the sauropods. The famous Brachiosaurus and the Diplodocus filled the fern prairies, feasting on the new tropical forests and habitats. The carnivorous Allosaurus also rose to prominence as it indulged on large herbivores such as the Stegosaurus and other sauropods. This ability to chow down on prey almost the same size as itself was facilitated by the ability of opening its jaw to a 92 degree angle and a bite force of up to 9,000 N. For reference, humans can only open their jaws up to 25 degrees with a bite force of 600 N.

As the Jurassic came to a close, the Mesozoic era entered its longest period yet; the Cretaceous, spanning 79 million years. During this time, volcanism began to rear its destructive head once again. As a result, the global climate became warmer and sea levels rose. This caused some portions of land to be flooded and was marked by the appearance of inland seas and a general shrinking of the continents as their margins flooded. Shockingly, almost half of the current land mass on Earth today was submerged during this time. Current values for the land volume state that it covers 28% of Earth's surface but in the Cretaceous this value had shrunk to 18%.

As far as plants were concerned, this was a pivotal time for flowers, or angiosperms. Previously, we talked about the emergence of gymnosperms during the Permian period. It can certainly be said that these were the first seed producing plants, but the key difference between the gymnosperms and the angiosperms (flowering plants) is how the seeds are developed.

The angiosperms are responsible for 90% of modern plant life, and develop their seeds in the ovaries of flowers which are then surrounded by the flesh of a fruit. These are the same seeds that you might encounter in any fruit, be it apple, orange, or watermelon. The gymnosperms however, lacked this ability to produce flowers and fruits, and had their seeds formed in cones. As a result, it was much harder for these plants to disperse their seeds far and wide, compared to the attractive nature of the fruits and flowers of the angiosperms.

In the seas, the role of the Ichthyosaurs as the apex predator had been overshadowed by a bigger, more formidable species. The Ichthyosaurs became obsolete during the middle of the Cretaceous, and the frightening Mosasaur rose up (swam up?) and took its place. As for our long-necked friends the Plesiosaurs, some species had gone extinct while others managed to live out the entirety of this period. Anatomically common organisms started to thrive during this time as sharks, rays, and finned fishes became more common and widespread.

Finally, we reach the culmination of our time here at camp Cretaceous. Our apex dino predators had reached the peak of their diversification. Just name it and it was most likely present during the Cretaceous. T-rex? Top of the food chain, rampaging through the land, gorging on any carrion and herbivores it found. Spinosaurus? Filled the void left by the decline of sauropods, further increasing competition amongst the apex predators. Triceratops? Feisty herbivores, using their horns and frills during intense battles with larger predators. Velociraptor? A feathered carnivore with that trademark enlarged sickle-shaped claw, used to pierce and slash vital organs of its prey. Quetzalcoatlus? The largest flying animal ever, a genus of pterosaur which had a wing span of up to 10 metres, and a height of 3.5 metres.

And so, the complete diorama reveals itself. The story of how these prehistoric beasts emerged from the devastation of the Permian extinction event on a ravaged ecosystem, its survival and success in colonising all domains of Earth, be it land, ocean, or sky, and its eventual demise during that fateful meteor impact. But the question still remains, why do we have such a special interest in these creatures that lived millions of years ago? What is our connection to them?

Exploring the Essence of Everything

Palaeontologist Mark Witton has expressed his thoughts on this matter, saying that dinosaurs combine and express 'high impact' anatomy. He explains this using a venn diagram, showing that dinosaurs have a combination of visual impact (in terms of size and proportions), exotic anatomy (such as the frills of a Triceratops, the teeth and jaws of a T-rex, or the armoured body of an Ankylosaurus), and familiar anatomy (body structures that are similar to modern day animals and life that society can visualise easily).

When we compare these features to other prehistoric life, we can see a definite answer where dinosaurs are clear of the competition. While the amniotes and arthropods of the early Proterozoic were anatomically exotic while being familiar to current ocean life, their small size is just not captivating to society, especially children. The large millipedes and dragonflies of the Permian certainly had intriguing sizes, never seen before on Earth, but their familiarity to current life prevents them from really providing an interest in the general history of life.

Dinosaur biology however, as Mark Whitton quotes, *"is thus near perfect for outreach material: they're visually impressive, anatomically and biologically accessible, but different enough to warrant interest"*. These creatures provide a gateway of interest to the layman, in discussions of evolution, adaptation, biological diversity, extinction, and the ever changing nature of the planet. Talk to anyone about the history of life on Earth and they will tune you out in a couple minutes. Introduce dinosaurs into the mix, and immediately curiosities are piqued and horizons are broadened. It is perhaps poetic that the greatest contributions these remarkable reptiles have made in history, is in the present day, introducing a world of science, history, and wonder, to the generation of tomorrow.

We have seen that the Proterozoic era belonged to the amphibians, the Mesozoic era was ruled by the reptiles, but what of the Cenozoic era? This is where we take the progressive step from past into present, as we explore the history of our own species.

History of Earth

Class Mammalia

Since we began this journey across time with some perspective from the Cosmic Calendar, let us end it in a similar manner. The current geological era, the Cenozoic, has spanned 66 million years, starting with the death of the dinosaurs, up to the present day. When we first embarked on this odyssey back in the Hadean eon, it was the 2nd of September. We then checked in 2 billion years later, on the 27th of October, the start of the Proterozoic eon. That was approximately 2.4 billion years ago. Up to this point, we have covered roughly 4.434 billion years of Earth's history. All that remains is the last 66 million years.

If 2.4 billion years ago was marked by October 27th, then at our current standpoint, we would be on the 30th of December, at precisely 6:24 am. Just 42 hours remain in our Cosmic Calendar. A story that started on the first second of January 1st with the Big Bang, is finally coming to a close. What can we hope to achieve during these final 42 hours? Well quite a bit actually. We still need mammals to complete their evolution, angiosperms to cover the surface of all land, the continents to drift to their current positions, and the emergence of a rather special group of organisms, *homo sapiens*.

The Cenozoic era ('new life'), started in similar fashion to its predecessor, the Mesozoic. The birth of new lifeforms after a catastrophic extinction event. Following the asteroid impact at the Chicxulub crater, in and on the coast of Mexico, 75% of all life had been wiped out, including all non-avian dinosaurs. Once again, it was time for a reset and for another group of organisms to rule the Earth, class Mammalia.

The first period of the Cenozoic, the Paleogene, spanned 43 million years, as the structure and life on Earth began to resemble modern day conditions. Firstly, Pangaea had completely broken off into its constituent continents. These continents, after millions of years of continental drift, finally settled into their current shapes and locations.

Exploring the Essence of Everything

The collision of India with Asia formed the Himalayas, South and North America were edging closer together, Australia had completely separated from Antarctica, and the Mediterranean sea started to appear between Africa and Europe. A movement that had started 4 billion years ago with the first landmasses that rose from the depths of the oceans had finally come to rest.

Don't be fooled though. The future of paleogeography indicates that we are due for another continental change. Of course, these changes are so gradual and miniscule that we rarely notice the effects year after year. But some future human civilisation or even alien civilisation will no doubt find the Earth much different than it is today, with some scientist predicting the formation of another supercontinent, Pangaea proxima, sometime within the next 250 million years. A blink of the cosmic eye.

In terms of the Cenozoic era being dubbed the Age of Mammals, this moniker was already in its adolescence during the Paleogene. In truth, the first ancestors of mammals began to emerge during the Cretaceous period back in the Mesozoic. A handful of these would have survived the Cretaceous extinction event, and continued to diversify and evolve, leading to the mammals of today.

As usual, we will explore all three domains of the Earth where life thrived. In the oceans, all marine reptilian life had gone extinct. The Ichthyosaur, the Plesiosaurs, the Mosasaurs. All gone, reduced to nothing but bone and ash etched in the sedimentary strata of the ocean floor. As a result, the seas were dominated by sharks, whales, and other cetaceans.

On land, a fairly familiar scene began to materialise. Vast jungles filled the surface, some even stretching as far as the poles. Creodonts, a now extinct order of carnivorous mammals, filled these jungles before being outcompeted by their evolutionary cousins, the Carnivora.

As the creodonts mainly roamed the jungle floor, the primates had started to colonise the tree tops, the earliest ancestors of modern day humans. This is where it all began for us. Swinging the treetops, munching on lice, fascinated by our bipedalism and opposable thumbs.

History of Earth

As previously stated, the lords of the skies were still very much dinosaurs in genealogy, if not in appearance. The birds took over the role left by the extinct pterosaurs as the last remains of the extinct dinosaurs. Over time, these winged vertebrates would evolve to include flightless species such as the penguin, ostrich, and emu.

The Neogene period is perhaps best known for the emergence of a rather intriguing and unique species amongst all others that have walked, swam, or flown on the Earth. *Homo habilis,* one of, if not the earliest form of anatomically modern humans, reared its head in Africa some 2.8 million years ago, with fossils found in Kenya, Tanzania, and Ethiopia.

These primitive humans, while not the same species as us, shared remarkable similarity with their primate ancestors. Most of *homo habilis* society was similar to chimpanzees and baboons, with which we share the highest percentage of our DNA, 98.9%. Some of these pre-humans even lived in trees, leading a partial arboreal lifestyle, which is shown by their long arms adapted for climbing and swinging.

Similar to primates, *homo habilis* was a scavenging species rather than a hunting one. They had omnivorous diets, consuming fish, fruits, and meat. The latter actually played a vital role in the growth of the brain compared to its australopithecus cousins. The hypothesis is that the energy and calorie dense protein allowed the ape gut to decrease in size, as more energy was obtained for less work. In addition, the ability to cook food rather than eating it raw, allowed for more efficient energy consumption and digestion, further decreasing the amount of space needed to break down food. This decrease in gut size allowed more energy to be devoted to brain growth, among other things. Refer to the diagram below for a visualisation of this anatomical change.

Technologically, *homo habilis* developed the use of the first stone tools. As brain size and subsequently cognitive thought increased, more

hurdles were overcome as this hardy species became the first problem solvers. Tools were used to butcher and skin animals, crush bones, and were even used to construct the oldest, primitive signs of human architecture.

While *homo habilis* continued to migrate and change the surrounding landscape, the Earth very quickly slipped into its current geological period, the Quaternary. Spanning the last 2.5 million years, this current period is not without its extinction events and development of new life. The Earth is a constantly changing myriad of life and death.

This period is marked by numerous ice ages, cycles of glaciation caused by the movement of ice sheets in a 40,000 to 100,000 year cycle. These glacial periods alternate with periods of warmer interglacial periods with the last glacial event ending 11,700 years ago. For those that have seen the animated movie *Ice Age,* many of the noticeable animals in the movie evolved during this period and have gone extinct over the past few thousand years. The sabre-toothed tiger, woolly mammoth, and glyptodonts all had rather short stays on this Earth, spanning just a few million years. But through the power of movie magic, these creatures have not been forgotten, and are instead used to wow the masses and invoke interest in children on these extinct species.

Let us continue the story of genus homo. A few hundred thousand years after the emergence of *homo habilis,* the Earth had become populated with more species of human-like ancestors. *Homo ergaster* and *homo erectus* start to colonise the African landmass and show more signs of migration. In fact, *homo erectus* is the first confirmed hominin to have lived outside of Africa, with fossils being found in Europe, China, and even as far as Java.

Homo erectus represented a quite large physiological jump for prehistoric hominids as their sizes were 60% larger than its predecessors. Additionally, it showed the greatest increase in brain size, owing to the munching meat hypothesis. Amongst the other hominids of its time, *homo erectus* is seen as a direct human ancestor.

Following this, the next significant life form in the line of genus homo was *homo heidelbergensis*. These hominids lived just 600,000 years ago, and are a crucial cog in this story, linking a few different species of hominids together. It is thought that *homo heidelbergensis* gave rise to modern humans, as well as our cousins, the Neanderthals and Denisovans. This species showed remarkable advancement, with evidence of stone-tipped and wooden spears being found in South Africa and Germany.

Further study of *homo heidelbergensis* fossils have indicated that these were the first hominins capable of speech, with the expression of the FOXP2 gene. However, this speech would have been mostly incomprehensible, as it was produced by balloon-like structures connected to the voice box, mainly used to form loud, booming noises.

The last and most recent species of hominids, *homo sapiens,* first arose 350,000 to 270,000 years ago. After a hundred thousand years or so in Africa, they start to migrate out, getting as far as the middle east. 70,000 years ago, the second migration of *homo sapiens* went global, reaching Europe, Australia, and Asia. A few thousand years later, our cousins, the Neanderthals, die out and *homo sapiens* are the last hominin species on Earth. However, there has been evidence that we met, lived with, fought, and even interbred with our Neanderthal cousins before their demise. Did we kill them off or was it a more natural end?

Exploring the Essence of Everything

The human lineage

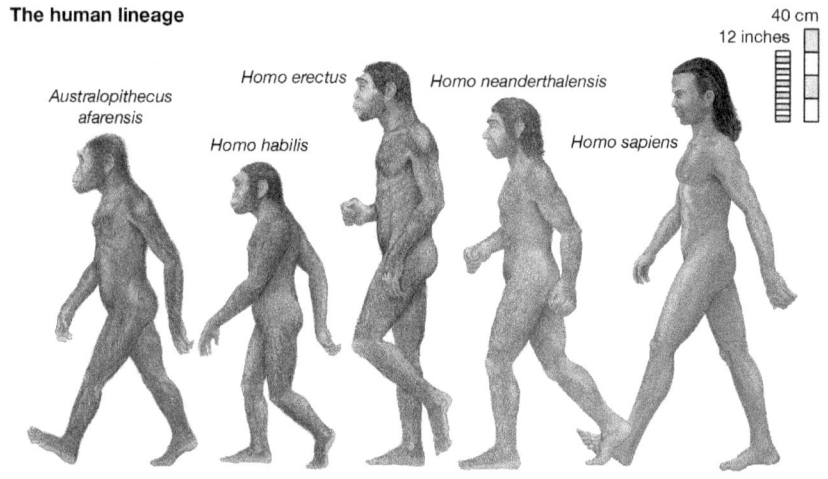

The past 11,700 years, known as the Holocene epoch, has accounted for much of human history. It has recorded the proliferation, growth, and impact of humans worldwide, including all of written history, ancient civilizations, technological revolutions, and urban industrialisations.

Some of these human activities have even started the next great extinction event, with families of mammals, reptiles, fungi, and other terrestrial wildlife being wiped out by widespread human activity. This has even caused the degradation of rainforests, coral reefs, and savannahs. It is said that the current rate of species extinction is 100 to 1,000 times faster than the natural background extinction rates.

At long last, we reach the zenith of evolution. 4.5 billion years of ever changing seas, airs, and landmasses. 4.5 billion years of biological evolution to reach this exact point in time. Stromatolites, arthropods, amphibians, fish, dinosaurs, birds, mammals. Are we the peak of evolutionary life forms? Or are we simply another step in this sequence of life? Will we even be remembered in another 4.5 billion years? I think not.

History of Earth

Throughout this chapter I have consistently used timelines and geological dates to represent the flow of time, as well as to provide a sense of scale. This rock was found 3.2 billion years ago, this fossil is 1.4 billion years old, this species was alive 4,000 years ago. But how do we really know all this? I hinted at it in the previous chapter, with the works of Becquerel and Curie. This is how we use chemistry, mathematics, and the natural state of matter to determine the age of everything, from the oldest stars in our universe, to the first fossils on Earth.

Radioactive Rock

For this next section, let us begin with a fascinating thought experiment. Imagine a box filled with atoms. Let's say 1,000 atoms. These atoms are in constant motion, neither trying to escape the box or trying to stay in it. They just exist. Whether any one atom escapes or not, is completely unpredictable. It is a random and spontaneous process. It could happen today, tomorrow, or a billion years from now.

However, we can use the law of averages and probability to determine how many atoms will escape from the box in a given time period. Let me explain. Let's assume that we can determine how long it will take half of the sample to leave the box. We don't know exactly when each atom leaves it, but we can extrapolate and form an aggregate for the total amount of time required. Let's say it takes 2 hours for the sample of 1,000 atoms to reduce by half. So now there are only 500 atoms in the box.

We need a bit more information before we can form a conclusive relationship. Alright, let us observe the box for another 2 hours. We find that in another 2 hours, the total number of atoms has reduced by half again. Only 250 atoms remain. And if we repeat this, 125 atoms remain after another 2 hour timespan. We see a pattern emerging now.

Using these three values; the initial number of atoms in the box (1,000), the time taken for half of the atoms to leave the box (2 hours), and the number of atoms left at the end of the experiment (125), we can determine how long a substance has existed for. This is the magic of half-lives.

In reality, the atoms escaping the box is just an analogy for alpha and beta emissions. These are spontaneous radioactive processes that can happen at any time in unstable atoms. But, we can determine how long it will take a sample of a radioactive substance to decay by half of its original quantity. This is called the half-life of an atom.

Another example, now that we have a grasp on the concept of half-lives. This time, we know how much the initial amount is, we know how much the final amount is, we know how long the total experiment takes, but we don't know the half-life. Let's give an initial value of 80 atoms, a final value of 5 atoms, and a total time taken of 20 days. First step is to keep dividing 80 in half until we get 5. 80 to 40, 40 to 20, 20 to 10, 10 to 5. A total of 4 divisions. This means that the atoms have gone through 4 half-lives. The total time taken was 20 days, meaning that one half-life of the atom takes 5 days. It takes 5 days for any sample of the atom to decay by half. As long as we have three of the four quantities needed, we can always calculate the final quantity.

In geology and carbon dating, the three quantities we have are the initial amount of the atom, the final amount, and its half-life which has been determined beforehand in laboratories. Remember, each atom has its own half-life which is constant. If one sample of cobalt-60 has a half life of 5.3 years, then every single sample of cobalt-60 will have the same half-life, regardless of initial or final amount.

So when geologists and archeologists need to determine the age of a fossil or rock, how do they do it? Let's use another thought experiment.

Imagine a time traveller. Our time traveller, lets name him Marty, has travelled back in time, and is now stuck in the past. He needs to know exactly how many years he has jumped back in order to return to the present. Bizarrely, Marty has a family heirloom that he has brought with him from the present. A tooth from the now extinct sabre-toothed tiger. It has a certain shape, size, and colour making it distinguishable

History of Earth

from other common teeth. Now, somehow, Marty is able to find the exact same tooth in the past. Same shape, size, and colour. It is undoubtedly the exact same tooth, just in the past.

So now Marty has two samples of the exact same tooth, one from the present and one from the past. Through some ingenuity and experiments, he is able to determine the percentage of carbon-14 in the present tooth, as well as the past tooth. The key here is that it must be the same tooth, which it is. Marty also knows that the half-life of carbon-14 is 5,730 years. Using these three pieces of information he will be able to determine how far back he has travelled.

Let's say that there is 40% of carbon-14 in the past tooth, and 1.25% of carbon-14 in the present tooth. This means that the carbon-14 in the tooth has gone through 5 half-lives. One half life takes 5,730 years. 5 half-lives will take 28,650 years. Marty now knows that he has travelled back 28,650 years, just by studying the change in percentage of carbon-14 in a tooth. This is carbon dating, the method used to determine the age of organic matter on Earth.

However, there is one catch to carbon-dating. It only works on timescales less than 50,000 years due to its short half-life. Any longer than that, and the amount of carbon-14 leftover will be insufficient to calculate an answer. Thus, for samples of organic matter older than 50,000 years, scientists use other unstable atoms such as potassium-40 with a half-life of 1.26 billion years, and beryllium-10 with a half-life of 1.52 million years. However, the principle is exactly the same. Compare the initial and final, and multiply the number of half-lives with the time taken for one half-life to find the actual age of the sample.

Although, there appears to be another hitch. Well a couple actually. Marty was able to figure out the initial amount of carbon-14 because he travelled back in time and found the original tooth sample. Obviously scientists don't have this technology. So how do they determine how much radioactive decay has occurred? Additionally, I gave examples where the half-lives of the unstable atoms were measurable, ie. 2 hours or a few days. What happens if the half-lives are impractical for us to measure? How do we know that carbon has a half life of 5,730 years or that potassium-40 has a half life of 1.26 billion years?

Let me answer the former question first. In determining the initial amount of radioactive substance in a sample, archaeologists have to use a fair bit of assumption, comparison, and estimation. Carbon-14 levels are actually quite simple enough to calculate. This is due to its main isotope, carbon-12. You see, throughout an organism's life, it takes in equal amounts of carbon-12 and carbon-14 in proportion. That is to say, for every 99 atoms of carbon-12, one atom of carbon-14 is inhaled. This ratio has been scaled up for simplicity's sake, but the key here is that the ratio never changes as long as the being is alive.

As soon as the organism dies and starts decomposing, the carbon-12 is locked in place. It is no longer taking in carbon-12 or 14 anymore. So whatever carbon-12 is left at time of death will remain so virtually forever, since carbon-12 is extremely stable and is not radioactive. On the other hand, carbon-14 levels will steadily decline due to its radioactive emissions.

So, all archaeologists have to do is measure the amount of carbon-12 locked in place, then use that value to estimate how much carbon-14 would have been in the organism when it died. This is the initial amount of carbon-14. By comparing this initial value to the present value, we can see how much the carbon-14 level has reduced, and using its half-life, can successfully calculate the age of the sample. It goes without saying that the sample must be organic in nature, capable of absorbing carbon-12 and 14 from the atmosphere. Carbon dating very rarely works on inorganic matter such as rocks and fossils.

For rocks and fossils, potassium-argon dating is often used. In principle, the concept is similar to that of carbon-12/14 dating. In this version, potassium-40, an unstable atom, will undergo positron decay to form argon-40. When studying a rock or fossil sample, two things are measured. The amount of potassium-40 remaining, and the amount of argon-40 accumulated. Once these are obtained, one just has to compare the amount of potassium-40 decayed with its half-life, 1.26 billion years, and the age of the sample can be found. In reality, this process is much more complex than I have just stated, involving a lot of assumptions and estimations, but the steps taken are the same.

Next, the matter of half lives. How can we possibly measure a half-life with a million or even billion year timeframe? One word, extrapolation. When calculating and analysing half-lives, scientists always sketch a graph of concentration (or amount) against time (half-lives).

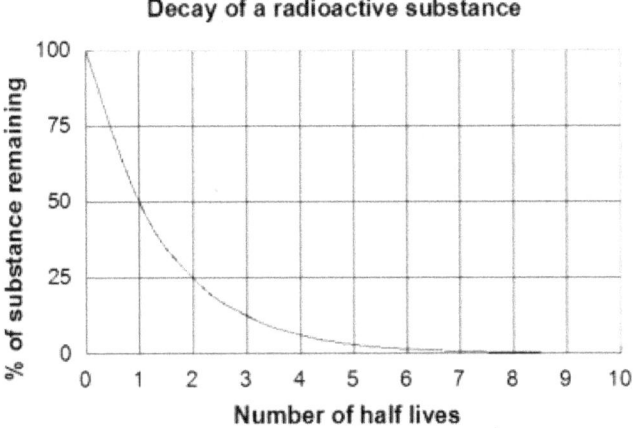

Imagine we have 100 billion potassium-40 atoms in a box. We know that potassium-40 has a half-life of 1.2 billion years, so it would take that amount of time for half the atoms to decay, leaving us with 50 billion atoms, and so on. Of course, we are not going to wait around for a billion years to verify whether this is actually true or not. So the next best thing? Instead of waiting for half to decay, why not just wait for a tiny, tiny fraction to decay? In this case, 0.000001%, or 1000 atoms. By finding out how long it takes these 1000 atoms to decay, we can then extrapolate this information to find out the half life of the entire sample. This would give us a timeframe of 12.6 years, and a half-life of 1.2 billion years. Much more doable.

However, since we're only relying on an incredibly tiny amount to decay, and assuming that this decay follows the general decay rate, our value for the half-life can vary greatly. Also, it is incredibly hard to put these values in a graph. The concentration/time graph for half-lives is best when measuring time in hundreds or even thousands of years. When we're talking about millions and billions of years, an equation is used instead:

$$N(t) = N_0 \left(\frac{1}{2}\right)^{\frac{t}{t_{1/2}}}$$

$N(t)$ = quantity of substance remaining
N_0 = initial quantity of substance
t = time elapsed
$t_{\frac{1}{2}}$ = half life of substance

At last, all our questions and queries concerning half-lives, carbon dating, and radioactive atoms have been answered. All that remains is to see how we apply these methods when writing the story of Earth. We will cover two types of samples that are measured using these methods; rocks and fossils.

Firstly, rocks. The oldest rocks on Earth date all the way back to the Hadean eon, 4.5 billion years ago. These rocks, in fact all rocks, usually consist of two or more minerals mixed up through geological processes. The oldest terrestrial mineral found in a rock sample is a zircon crystal, with an age of 4.4 billion years. Scientists found this rock in the Jack Hills valley of Western Australia, and determined the isotopic ratio in the zircon to calculate its age.

That being said, zircon is only the oldest terrestrial rock on Earth. There have been instances where non-terrestrial rock samples have recorded far older ages, and have even been used to determine the formation of the moon.

History of Earth

In 1971, during the Apollo 14 mission to the Moon, astronauts brought back a lunar rock sample called Big Bertha, named after the German WW I howitzer. 48 years later, in January 2019, analysis of the rock showed that minerals such as quartz and granite, common to the Earth but uncommon to the moon, were found in the rock. This proved, or at least made it very likely, that Bertha was formerly a part of Earth, flung apart during the collision with Theia, and went on to form a piece of the lunar surface. By studying the zircon crystals in the rock, an age of 4 billion years was determined, making Bertha one of the oldest known Earth rocks.

The inverse has also happened. Just as fragments of Earth can be detected in the solar system, and indeed brought back, so can foreign pieces of the universe be found on the Earth's surface. The Murchison meteorite is one such example.

The Murchison meteorite struck the village of Murchison in Victoria, Australia, on September 28th, 1969. Perhaps there is some magnet under Australia that can attract all these ancient rocks and minerals. Weighing over 100 kg, this rock sample is confirmed to be the oldest material ever found on Earth, with a staggering age of 7 billion years. This predates the age of our very own solar system and Sun by a mere 2.5 billion years! Imagine, this meteorite has been traversing the universe even before the first fusion reactions in our Sun, before the collapse of the nebula which formed our solar system, probably even before the death of the star from whence we all came.

The most fascinating thing about the Murchison meteorite is not even its age. It's the fact that this piece of rock, some 7 billion years old, actually contains organic matter and chemicals that are used to form life here on Earth! This lends further credence to the hypothesis that the building blocks of life were brought to our planet through one, or many, of the asteroid collisions during the Hadean eon, a theory called panspermia.

Staying on the topic of rocks, we encounter a different sample; sedimentary rock. Sedimentary rock is formed when organic matter and

Exploring the Essence of Everything

minerals deposit and accumulate on the Earth's surface, over millions of years. These sediments, hence the name, are deposited in thin layers called strata, a sort of snapshot in the history of the rock's formation.

Probably the most notable form of sedimentary rock are fossils. Here's a fun fact on fossils. Whenever you go to a museum, and marvel at the bones and skeletons of so many prehistoric dinosaurs, you're not actually looking at the bones of the dinosaurs themselves. The bones of the dead dinosaurs would have disintegrated long, long before humans even emerged. What palaeontologists actually dig up, are the fossilised remains of the animal bones. What happens is, the bones will deteriorate leaving behind a cast. This cast is then filled with sediments, which, after millions of years, harden to form sedimentary rock. These rocks have essentially been moulded into the exact shape of the dinosaur skeleton that came before it. These fossils are so exact a recreation of the original bones, that scientists can determine an organism's diet, where it lived, how its species evolved, and even how it died. It is these fossils that are dug up, cleaned, analysed, and then displayed for public admiration.

Of course, fossils are not exclusive to just the dinosaurs. Another example of the superiority of these animals over any other life form on the Earth. As soon as we hear 'fossils' we immediately think of the T-rex and Brachiosaurus. In reality, fossils can be formed from any type of organic matter.

The oldest fossils on Earth are familiar to us. We have already covered them during our journey through the history of life on Earth. Of course, I refer to the stromatolites of cyanobacteria, reaching up to 3.8 billion years in age. Apart from this, microfossils such as the haematite tubes from hydrothermal vents, are equally as old. Formed from the iron ore haematite, these tiny, microscopic fossils could have been breeding grounds for primaeval bacteria.

Alas, we reach the end of our exploration of Earth's rock and fossil record. We now have a much better understanding of the concept of half-lives, how they are used to calculate the age of a rock, fossil, or organic matter, and how we can determine even the longest of half-lives.

History of Earth

I suppose our study on Earth and its history as a whole has reached its conclusion. From the hellish Hadean, to the vistas of the Proterozoic, to the teeming lifeforms of the Phanerozoic, we have shown the progression of the Earth and life through much adversity, multiple extinction events, and so much trial and tribulation. What we can conclude, is that both the Earth and its life seem to have this knack of constantly surviving and thriving in whatsoever condition. This Earth of ours, and indeed its life, has been here long before us, and it will be here long after us. Perhaps there is something poetic in that. We are merely a brief pitstop in the story of this extraordinary rock.

Be that as it may, this is not the end of our analysis on all things terran. In fact, now we will see what actually makes the Earth tick. Let us answer the all important question. Why Earth? Why not Venus or Mars or Jupiter? What is so special about this Earth of ours, that it seems to be the only hospitable place (that we know of) in the entire universe? Are planets like ours common throughout the universe, like so many stars in the cosmos? Or are we simply...

One in a Septillion

We will conclude this chapter with a bit of history which was all the rage back in the 16th century. Before Einstein, Hubble, Newton, and Kepler, there was Copernicus. This polyglot and polymath, living in Royal Prussia (modern day Poland) during the height of the renaissance, changed the way people and scientists of the day thought about the location of Earth in the grand scheme of the universe. Unfortunately for Copernicus, his ideas and theories also challenged the judge, jury, and executioner of the time, the Catholic church.

We touched on this briefly during our tales of Kepler back in chapter 1, but let us go through it in more detail. During this time, the geocentric model of the universe was popular amongst scholars and popes alike, due to the fact that it placed the Earth at the centre of the universe. You can imagine what gratification certain religious societies would have received from this stipulation. The Earth at the centre of an all-knowing universe, a position of immense power and control. This could only have been fashioned by an Almighty hand, building the universe around us for our viewing pleasure.

Whatever your thoughts on these ideas, it was proved false by Copernicus and subsequent scientists after him. Copernicus, after rigorous experimentation and thought, brought forward the concept that the Sun, not the Earth, was the true centre of the universe, and that all other celestial bodies revolved around it. This was altered in the following centuries as it became clear that the Sun is just the centre of our solar system, while the universe as a whole does not have a true centre, just points of reference.

I'm sure you can see how this model upset a pious and god-fearing society. From its special, privileged place at the centre of the universe, the Earth had been demoted to a mere dot in the universe, revolving around the true centre, just like all the other dots. The Earth had become ordinary and plain, clearly no deity on Earth or in the heavens above would have created a world so common and unremarkable? To this day, while the truth about the centre of the universe has long been settled, the idea that the Earth is just another unimportant rock still leaves a bitter taste for many. Surely we must have something unique that no other planet in the universe has? And indeed we do. We have life.

Now, whether there is alien life on other planets is not to be discussed in the scope of this chapter. For that, refer to Vol.3, Exoplanet Exploration and the Quest for Life. This chapter focuses all attention on the Earth. Its history and its uniqueness. Life may or may not be possible on other planets, but for now, we will simply talk about why it is possible on our planet.

When we speak about the uniqueness of Earth, we have to compare it with the other star-orbiting rocks that inhabit the cosmos. This comparison is done to provide a sense of how truly lucky we are, that

History of Earth

certain things worked out the way they did, leading to our eventual emergence and evolution.

The best estimate from astronomers and scientists currently, is that the universe has a septillion planets in it. To say that a septillion is a rather large number, is somewhat understating it. To visualise, a septillion is 1 followed by 24 zeros, or 1,000,000,000,000,000,000,000,000. Out of all the planets that make up this number, we account for just one. But how do we arrive at such a number? How do we even begin calculating this?

Let's work it out step by step. Firstly, we start off with our galaxy, the Milky Way. This is a fairly average-sized galaxy in the scale of the universe. Definitely bigger than a few, definitely smaller than the rest. Our galaxy has 100 billion stars in it. Again, we take this value as an average, saying that every galaxy has 100 billion stars. Of course, some will have much less and some will have much more. Observations are so potentially vast that we just have to take an average.

Next, we deduce the number of galaxies in the entire observable universe, which stands at 2 trillion galaxies. By multiplying these two numbers together, we get a total value of 200 billion trillion, or 200 sextillion stars (2 followed by 23 zeros).

The number of stars in the universe is actually much easier to calculate than the number of planets. We know what it takes to form a star. We know the average rate of star formation. We know the lifespans of different star types. A planet is slightly different.

We know that planets are formed from the debris of stars. Whatever is leftover that doesn't go into the star, forms the planets. However, the exact number of planets that can possibly orbit a star is highly unknown and not constant. Current estimates say that there is at least one planet for each star in the universe. But then again, our own star has 8 planets. The nearest star to us, Proxima Centauri, has three planets. The TRAPPIST-1 system has 7 planets, most of which are possible sources of life. Keep in mind that the exoplanets observed are the bigger, more noticeable ones. There could easily be Mars or Mercury-sized planets in these systems, increasing the total number of planets.

So we see that the minimum number of planets is 1, but the maximum number is anywhere from 8 to 12 potentially. This gives us an average of about 5 planets per star. Multiplying the total number of stars in the universe, 200 sextillion, with 5, gives us 1 septillion planets.

To give this number further context, I will refer to a saying popularised by our beloved Carl Sagan. Carl said that *"the total number of stars in the universe is larger than all the grains of sand on all the beaches of the planet Earth"*. A rough estimate has been provided by scientists, that there are 7.5 sextillion grains of sand on the Earth. If we compare this to the septillion planets in our universe, we would need 133 times the total number of grains of sand, to equal the total number of planets. Imagine every single grain of sand on Earth. Multiply this by 133. Now you have every single planet in the universe. From all these innumerable grains, we are one. How can we possibly be special? Is there even a scale for speciality when there are so many planets like ours? What sets us apart from the rest? Let us finally see the conditions that have made the Earth habitable for life.

We will divide these conditions into two sections. The first section will be for characteristics of the Earth in relation to its position and orientation in space, such as its tilt, its distance from the Sun, the presence of the moon, its circular orbit and so on. The second section shall contain characteristics of the Earth that are bound to its surface and interior, such as its atmosphere, its magnetic field, its core, etc.

Let's start with the Earth's position in space. Numerically, our planet is 26,000 light years away from the galactic centre. Quoting a number this large doesn't really do much unless we give it some context. Since a light year is the distance that a photon of light can travel in a year, if we were to observe the galactic centre right now, we would see it as it was 26,000 years ago. The light that we see would have originated at a time where humans built huts from mammoth bones and slept with a blanket of sabre-tooth tiger fur.

History of Earth

Alright, so what if Earth is such a distance away from the centre of the galaxy? Surely the distance is so vast as to not affect us at all? It turns out that we are in a 'galactic sweetspot'. We are almost exactly in between the centre and the edge of the galaxy. This matters due to three reasons.

The first, puts an outer limit on the distance from the centre. Why a planet or star system can't be too far away. It states that as distance from the centre increases the metallicity of stars decreases. Remember that metallicity of stars is essentially all elements apart from hydrogen and helium. As we get further out towards the edge of our galaxy, the stars contain very little metal, meaning that it would be increasingly difficult to form terrestrial planets. The Earth for example, is made up of more than 50% metal.

The second and third reasons put an inner limit on the distance from the galactic centre. Why a planet or star system can't be too close. The first is X-ray and gamma ray radiation from the supermassive blackhole at the centre of the galaxy, around which all celestial objects revolve. These highly ionising rays can have damaging effects on any life that arises on a planet that is too close to the centre.

The final reason is the gravitational perturbations and disturbance of planets and stars by other nearby stars and black holes. Imagine the New York City subway during rush hour. Hordes of people bumping and crashing into each other, barely any space to breathe. Now imagine a quiet suburb, perhaps the Catskills. Low density of people, low density of cars, less chance of being knocked down. And so the same is true for planets. Any objects near the centre stand a far greater chance of being smashed into by a rogue asteroid or even ejected completely from its star system due to the pull of an even larger star.

When we talk about the distance of Earth from the Sun, it is for far different reasons. Our distance from the Sun influences two vital aspects of life. The temperature, and the presence of liquid water. And indeed, both these things are inextricably linked.

Exploring the Essence of Everything

In astrophysical terminology, there is such a thing called the 'Goldilocks zone'. This refers to a situation, where the surface temperature of a planet, much like an ideal bowl of porridge, is not too hot, not too cold, but just right. This is a fairly simple principle to understand, the planet Neptune for example, is 30 astronomical units, or 30 times the distance between the Earth and the Sun. As a result, temperatures on Neptune can fall as low as -218°C. On the other hand, Mercury, a mere 0.4 astronomical units away from the Sun, has temperatures as high as 430°C. Liquid water wouldn't be possible on either of these planets, in its natural state. And before you say that Venus is actually the hottest planet in our solar system, granted that is true, Venus's scalding temperature is due to greenhouse gases in the atmosphere more so than its distance from the Sun. Perhaps our Earth is heading in a similar direction.

So we see that the Earth is truly in the sweet spot of space. Perfectly in between the radius of the galaxy, and at the exact distance away from the Sun to sustain liquid water. However, this will not be the case forever, as the Sun inevitably swells up after 5 billion more years of fusion, engulfing Mercury and Venus, and scorching the Earth. But that is a story for another time.

Next, we must discuss Earth's orientation in space. In this, we talk about three features; its tilt, its spin, and its circular orbit. The Earth is tilted at a 23.4° angle, which is not uncommon for our neighbouring planets. In fact, all the planets (except Mercury) are tilted to some degree, with the most extreme being Uranus, with a 82° tilt. Uranus actually rotates vertically rather than sideways like all the other planets, and any potential life there would experience drastic seasonal variations in climate. It is this tilt which gives us the four seasons. Without it, there is the argument that heat from the Sun would be unevenly spread, with some areas experiencing extreme temperatures while others get barely any heat at all. Once again, Earth finds itself in the sweet spot, or as the Swedes would say, 'lagom'.

History of Earth

For a more extreme situation, we can imagine what the Earth would be like if it had no spin. If the Earth was tidally locked, or didn't rotate, then one side would forever face the Sun, while the other side would be in a perpetual winter. Any possible life that evolved would be faced with rather harsh conditions, trapped between death by boiling or death by freezing.

The third spatial orientation also has to do with varying temperatures, this time on a longer timeframe. I mentioned this back in the first chapter, that the Earth has a relatively stable circular orbit. In fact, almost all the planets do, a peculiarity about our star system. At its closest point to the Sun (perihelion), the Earth is 147.1 million km away, while at its farthest distance (aphelion) it is 152.1 million km away. A difference of just 5 million km, a hair's width in the cosmic scale.

If the Earth had a more elliptical orbit, caused by larger planets closer to the Sun, temperatures could vary drastically. This would allow any life on Earth no chance whatsoever to get acclimated to its environment. If it somehow mutated the ability to withstand the hot conditions, it would be wiped out as the Earth travelled further away from the sun, and vice versa.

The last factor of space on the Earth has to do with its surrounding celestial bodies, mainly the Sun and the Moon. The effect of the Sun in this context is more indirect as opposed to its direct influence on temperature. In this circumstance, it is the age of the Sun, with which we are concerned. As mentioned before, the Sun is a G type star, with a lifespan of 10 billion years. We are just under halfway of its complete lifespan. The Sun is probably going through its mid-life crisis right about now. In the 4.5 billion years of the Sun's life, life on Earth has been allowed to emerge, evolve, and become intelligent (still highly debatable).

If the Sun were an O, B, or even A type star, it would have died long before life on Earth could even have a chance of evolving. Best case, all there would be is a handful of cyanobacteria, and even that is doubtful. An E type star would still exist now, but it would only have 500 million odd years left on its ticker. Enough time to revolutionise interstellar space travel?

The Moon has a different effect on Earth. While it does help to stabilise Earth's axial tilt, the effects of which we have already discussed, its main importance is its gravitational effects through tidal forces. These forces are further helped by the size of our Moon. In comparison with the other natural satellites, our Moon is the fifth largest, a fact which has not gone unnoticed.

While the obvious examples are the varying tides we experience daily, these forces played a crucial role in the formation of Earth's continental crust. The large moon increases the chances of plate tectonics on our planet, causing the crust to emerge from the ocean, and the continents to take shape. Without these effects, we would probably be an ocean-bound species, forever trapped under a veil of water.

The aforementioned factors for life are all space related, the Earth has no real say in them. They just are. These next few features are of the Earth itself, with little influence from outside sources. The main factors are; the atmosphere, the magnetic field, and mantle radiation.

Let's start from the centre and work our way out. We know that Earth's core consists of two layers. An inner solid mass of iron, and an outer layer of molten iron. There is also a sprinkle of nickel, 10% to be exact. This forms an iron-nickel alloy.

Firstly, heat from the inner core (5,600°C) causes the molten iron around it to move and flow. Eddy currents in the fluid, the convection of the fluid, and the Coriolis effect (force found in a rotating object), all combine to produce Earth's magnetic field. This magnetic field has a strength of 25 Gauss, 50 times stronger than the magnetic field on the surface. In truth, the actual process is slightly more complex, but as a general explanation, this will suffice.

This magnetic field has a very important role. It basically acts as a shield against the Sun. But a shield against what? The Sun ejects an absurd amount of highly charged particles from its surface per second, in the form of solar wind and coronal mass ejections. These charged particles could do irreparable damage to life on Earth, or any life trying to rear its primitive head. Not only would DNA be highly ionised, but the atmosphere itself could be stripped away. Mars is one such example of a planet without a magnetic field, naked and unprotected from all the dangers of space.

So our magnetic field repels, traps, and diverts these particles around the Earth. Some of the particles that react with elements in the upper atmosphere can cause the dazzling sights of the aurora borealis, while most of them form the doughnut-shaped zones called the Van Allen radiation belt.

Moving out of the core, we encounter Earth's thickest layer; the mantle. The mantle is unique due to the presence of radioactive isotopes embedded in it. There are four main radioisotopes at work here; uranium-232, uranium-235, thorium-232, and potassium-40. We know by now that these decay reactions release immense quantities of heat and energy. In fact, about 50% of Earth's internal heat comes from the decay of these unstable atoms. How does this affect life on Earth?

This internal heat caused two things. Volcanism and plate tectonics, and hydrothermal vents. We know that volcanism and plate tectonics were vital in creating the continental crust, as heat convection from the mantle caused the surface to rise from the depths of the ocean. Hydrothermal vents might seem foreign to us, but in our search for the first life forms on Earth, they are highly important. It is said that one of the places where life could have originated is in these hydrothermal vents. The exact circumstances and factors are explained in Vol.2, but for now it is adequate to say that these heated fissures on the seabed could not have occurred without internal heating from radioactive decay in the mantle.

Stay with me here. We are at our final factor. Rising from the inner regions of the Earth, we step forth into the atmosphere. While Earth's atmosphere can be separated into a handful of layers, for this purpose we are only concerned about one layer and one property. The ozone layer, and the pressure of the atmosphere. The existence of an atmosphere in the first place is mainly due to Earth's size and gravitational pull.

By now, we are familiar with the concept of pressure and its role in sustaining liquid water on the surface. The ozone layer is crucial for blocking and absorbing UV radiation from the Sun from reaching the surface. We clearly see that while the Sun is by far our best source of energy, this is not without its pitfalls, with abundant UV rays and charged particles barraging the Earth. We give thanks to our magnetic field and ozone layer for keeping these rays and particles at arms length and allowing life to flourish. However, other planets are not as fortunate.

Once again, we will use Mars as an apt example. Our red neighbour, with its low gravity and small size, is not able to hold a sufficient atmosphere. Whatever gases present on the surface account for just 2% of Earth's atmospheric density, while its pressure is less than 1% of the Earth's. An extremely thin and barren atmosphere, so to speak.

As mentioned before, even though the temperatures on Mars are well below the boiling point of water, the reason liquid water is not possible on the surface is due to this incredibly low atmospheric pressure. It is simply too easy for the molecules to evaporate into the air, to be dispersed across the planet. Another property of the Martian air is the lack of oxygen. While oxygen makes up 21% of air on Earth, it only accounts for just 0.2% of martian air. As a result, Mars has virtually no ozone layer, and any UV rays from the sun can be incredibly damaging to organic molecules on the surface.

And scene. At the end of a whirlwind adventure, I hope, dear reader, that you are now better acquainted with the miraculous circumstances that have brought you to this moment in time and space.

We have taken on the physics of our universe, and challenged our position in it. Where are we? How far away are we from everything else? What is the great circle of life, and how are we a part of it? These were the astrophysical and planetary science elements of our journey.

We then questioned the fundamental composition of the universe, and everything in it. How are we all connected? Are there invisible strands between us and the fabric of spacetime at large? What are the forces that hold us together? How did everything in the universe come to be? Will our particles ultimately be returned to the vast nebulae of the cosmos? Perhaps we are just the progenitors of future life in the universe. These were the astrochemical elements of our journey.

And finally, we learnt about the wonders of our home planet. How can we chronologize time on Earth? How can we pick up a rogue rock and gauge its age and place in history? What were the conditions that made life possible, and continue to do so? What does the flow of life on Earth look like, and how did it ultimately end with the emergence of *homo sapiens?* These were the geological and anthropological elements of our journey.

And at our crescendo, we find that this is merely a quarter of the complete journey. We are yet to brave the foreign worlds of distant star systems, encounter alien life, and debate the ultimate fate of the universe. However, before we can discuss the potential for life on exoplanets, we must first understand how life arose on our planet.

Join me in the next volume, as we ponder the very meaning of existence, and the idea that all life on Earth is connected through billions of years of evolution. How are we descendants of every single life form that has preceded us? How can life survive even in the harshest of conditions? What exactly does Darwin's theory state? What are the building blocks of life? What is the science of mutation and genetic variation? Natural selection, fact or fiction? These will be the biological elements of our journey.

I thank you for taking the time to follow me on this journey, exploring the essence of everything. We will rendezvous in Vol.2, where we will understand the essence of existence. Till next time...

James Hutton established geology as a modern science as early as 1785. Hutton pioneered research in fields such as rock formations, meteorology, and the age of the Earth. He published a number of books and papers over a span of ten years, and was a member of the Royal Society of Edinburgh.
(Henry Raeburn)

Arthur Holmes was an English geologist who performed the first accurate radiometric dating of a rock sample nearly 400 million years old. Holmes also formulated the modern geological time scale in 1911.
(The Geological Society)

Charles Lyell brought the idea of geology to the masses with his book *Principles of Geology*, a three volume text published between 1830 and 1833. He also provided valuable contributions to the research of volcanoes, glaciers, and fossils. (George J. Stodart)

Inge Lehmann was a Danish geophysicist, known for her work in the field of seismology. By studying earthquakes and seismic waves, she was able to deduce that the Earth contained four distinct layers. (Evan Neuhaus)

Preston Cloud was an American Earth scientist and was the first person to recognize the importance of Cambrian rocks and fossils in an evolutionary context. He proposed the idea of a 'Cambrian explosion', an event that caused rapid diversification of life on Earth. Over his lifetime, Cloud published over 200 works and papers, ranging from cosmology to geology. (University of California)

Arthur Philemon Coleman was a Canadian geologist who led multiple field expeditions to explore the natural formations around Canada. During a survey of Lake Huron, Coleman analysed geological formations and deduced the possibility of a glaciation event, occurring 2.4 million years ago. He has a glacier, lake, and mountain summit named after him. (University of Toronto)

Abraham Ortelius (top) was a cartographer in the 16th century. He is recognized for creating the first modern atlas, as well as proposing the idea of continental drift to explain the similarity between coastlines of different continents. Alfred Wegener (bottom) continued this research more than 300 years later by publishing his complete hypothesis on continental drift in 1912. This work included the science behind the process, something that Ortelius was not able to work out due to the limitations of the time.
Top - (Peter Paul Rubens)
Bottom - (University of Hamburg)

A stromatolite (top) found in Strelley Pool, Western Australia. This sample is more than 3.43 billion years old, making it one of the oldest fossils on Earth. These were the habitats and breeding grounds of cyanobacteria. (Fossilera)

Towers of stromatolites found at Pavilion Lake, British Columbia. These structures are the largest sample of freshwater stromatolites on Earth, and NASA has even conducted xenobiology research here. (Donnie Reid)

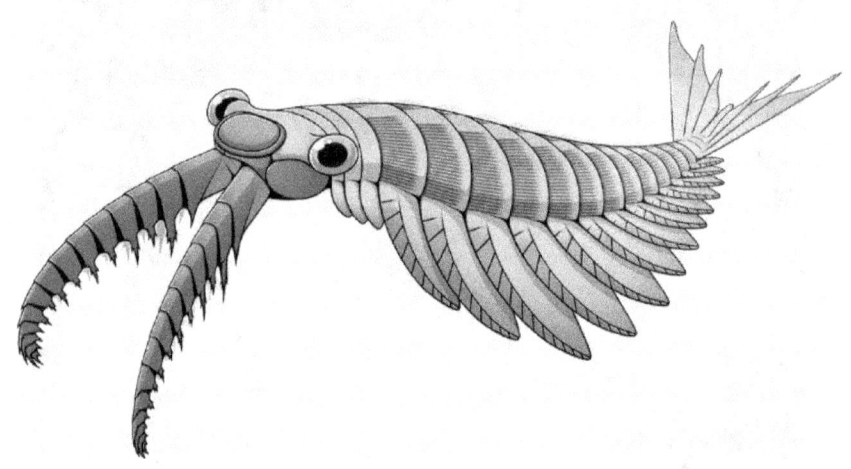

The *anomalocaris* was an early marine predator, one of the first arthropods to evolve in the oceans during the Cambrian explosion. This creature was estimated to be up to 37 cm long, with one of the most powerful eyes of any contemporary species.
(Junnn11)

The *plectronoceras sp.* is the earliest known shelled cephalopod, a member of the mollusc class. They were prominent during the Late Cambrian, as life continued to diversify. (Entelognathus)

An artist's depiction of Ordovian life in the oceans. This image, showing a scene some 450 million years ago, illustrates how far life had come from the miniscule microbes just a few million years ago.
(John Sibbick)

An artist's rendering of a trilobite, a sea arthropod that was commonplace during the Ordovician period. This cockroach-like creature is an ancestor of today's crustaceans, and went extinct 250 million years ago.
(Nobu Tamura)

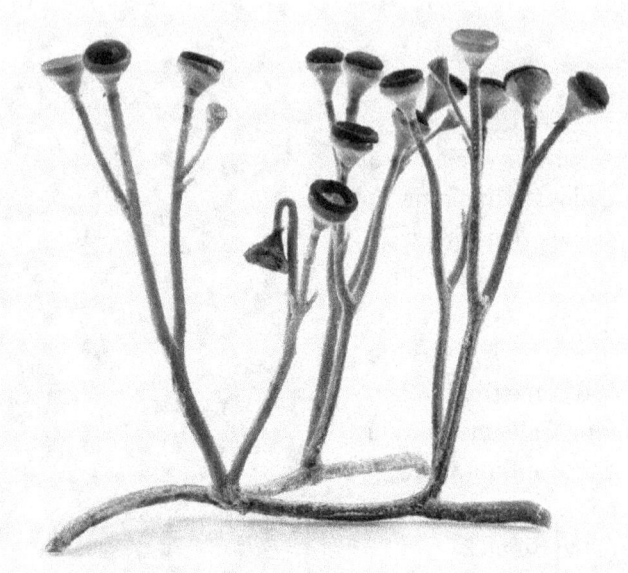

Cooksonia was one of the first forms of terrestrial plant life. It is the oldest plant to have a stem with vascular tissue for the transport of water. (Matteo de Stefano/MUSE)

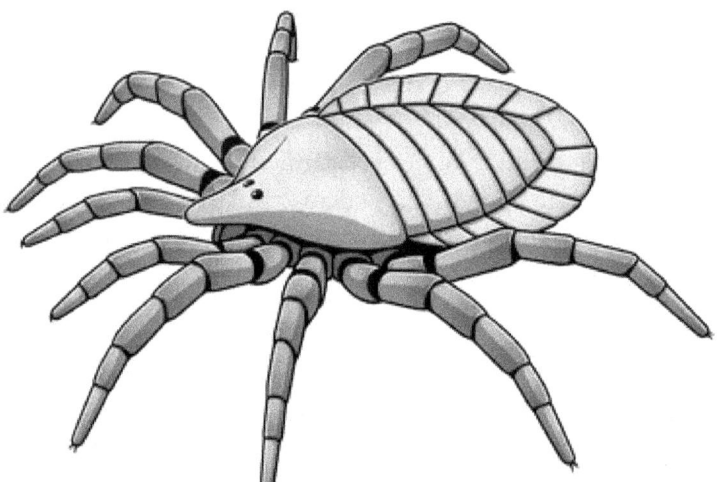

The *Trigonotarbida* order represented a new form of terrestrial life. These extinct arachnids emerged on land during the Silurian period and are the ancestors of modern day spiders.
(Junnn11)

An artist's interpretation of marine life during the Devonian period. This 'Age of Fishes' represented mass diversification as armoured fish began dominating the seas and rivers. The first ancestors to sharks emerged during this time.
(Gess Ahlberg)

The *tiktaalik* was one of the first tetrapods, considered to be the link between sea-based fish and land-based amphibians. These organisms' fins would eventually evolve to become the limbs of land animals.
(Zina Deretsky)

Etchings depicting what the trees and forests of the Carboniferous period looked like. Note the massive trees, some growing up to 30 metres in height, showing a remarkable transformation from the primitive *cooksonia*. The gymnosperms evolved during this time, plants with exposed seeds such as conifers and ginkgos. Their ancestors can be seen in these images.

Left - (Bibliographisches Institut)
Bottom - (Nature)

The *moschops capensis* is an example of a synapsid, one of the two forms of terrestrial vertebrates. The synapsids were the dominant land animals during the Late Paleozoic, and are known as the first ancestors of mammals. You can see how the limbs had evolved from the creatures that emerged from the seas, with hindlimbs and forelimbs now noticeable.
(Nobu Tamura)

The *hylonomous lyelli* was an early form of reptile, prominent during the Late Carboniferous. This organism is an example of the second type of terrestrial vertebrate during this time, the sauropsids. The sauropsids are the ancestors of modern day birds and reptiles.
(Nobu Tamura)

Gorgonopsia were carnivorous therapsids prominent during the Permian period. These creatures had evolved from the amphibians of the Carboniferous and are reminiscent of modern-day wolves. (Dmitry Bogdanov)

Due to an increase in oxygen levels, the size of some animals increased tenfold. Shown below are the *meganeura*, a dragonfly-like insect with a wingspan of 75 cm and the *arthropleura*, a millipede-like arthropod with a length of up to 2.6 metres.
(Owlcation)

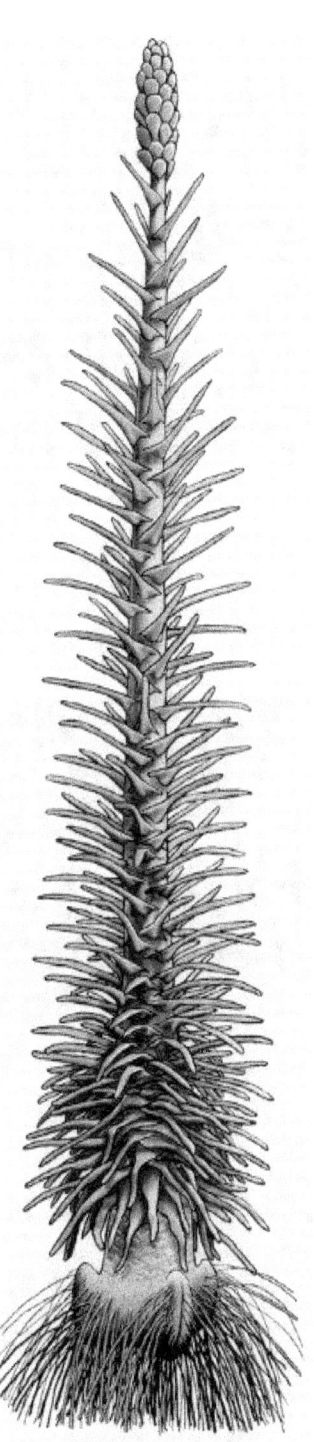

Pleuromeia dominated vegetation during the Early Triassic, right after the Permian extinction event. It could grow up to 2 metres in height, and thrived in a carbon dioxide rich atmosphere.
(Ivo Duijnstee/Hannah Bonner)

The *coelophysis* was one of the first types of dinosaurs to evolve. Its fossils have been found in New Mexico, and it was the second dinosaur to be sent to space, when a fossil of its skull travelled aboard the Space Shuttle Endeavour in 1998 and even boarded the Russian Space Station, Mir.
(Dr. Jeff Martz)

The *ichthyosaur* was the apex predator of the seas during the Triassic period. This fearsome carnivore evolved from prehistoric reptiles and looks like a hybrid of a dolphin and a crocodile.
(Dotted Yeti)

The *plesiosaur* was a marine reptile that has been described as "a snake threaded through the shell of a turtle". It could grow up to 15 feet in length and some conspiracy theorists absurdly say that the famous 'Loch-ness monster' is actually a *plesiosaur*.
(Roger Harris)

The *pterosaur* was the first vertebrate capable of flight, and had a diet mainly consisting of fish. This avian died out with the rest of the terrestrial dinosaurs and could have originally evolved from smaller, ground-dwelling reptiles. (Science News)

The *archaeopteryx* was the avian dinosaur and is even a contender for the title of 'first bird'. This ancestor of all modern-day birds was a cross between bird and terrestrial dinos as it had feathered wings and claws but also sharp teeth and a bony tail. (Bridgeman Images)

The *stegosaurus*, one of the most iconic dinosaurs, is instantly recognizable by the base plates on its spine. These herbivorous dinos emerged during the Jurassic period and died out well before the extinction even at the end of the Cretaceous.
(Nobu Tamura)

The *brachiosaurus* has been estimated to have a length of 21 metres and a height of 12 metres. This colossal herbivore dined exclusively on the tree tops and has fossils found in Africa, Europe, and North America.
(Dr. Scott Hartman)

The *allosaurus* was among the earliest dinosaur discoveries and was the apex predator of its time. This carnivore feasted on prey such as the *stegosaurus* and *apatosaurus*.
(Jean-Michel Girard)

The *ankylosaurus* is easily identified by its armoured body and club tail, used to protect itself from larger predators. The only way for it to be maimed was to flip it over, revealing a softer underbelly.
(Live Science)

During the early Cretaceous, the dominant predator in the oceans were the *mosasaurs*. These marine reptiles could grow up to 50 feet, with the preserved remains of their stomachs showing signs of eaten sharks, birds, and other *mosasaurs*.
(Thomas Miller)

The apex predator of the entire Mesozoic era, and perhaps the most famous dinosaur of all time, the *tyrannosaurus rex*. With dimensions of 40 feet in height and 20 feet in length, these carnivores simply gorged on any prey that happened to be in its way.
(Nature)

The *velociraptor* was actually something of a flightless bird, and differs from its depiction in movies. These dinosaurs actually had feathers and a long tail, with its trademark sickle-shaped claw visible.
(Fred Wierum)

The *quetzalcoatlus* was the largest flying creature of all time, with a wingspan of up to 10 metres. It had the ability to turn its head 180° and hunted small vertebrates on land.
(Johnson Mortimer)

The *triceratops* could weigh up to 10 tons and fed on ferns and fibrous plant material. It had up to 800 teeth as well as three horns and a large frill, making this one of the most distinct dinosaurs.
(Leonello Calvetti)

The *spinosaurus* was one of the rival predators of the *t-rex* during the Cretaceous period. It is the longest known terrestrial carnivore with a length of 14 metres, and resembles modern-day crocodiles.
(Nature)

The *creodonts* were the dominant carnivorous mammals during the Paleogene period and are some of the first forms of mammalian life to evolve after the Cretaceous extinction event. Many of the *creodonts* are direct ancestors of modern day predators.
(Apokryltaros)

Primates first evolved during the Paleogene period, 60 million years ago. Their long arms and slender bodies helped them climb and swing on trees to reach high-hanging fruit. These were the first organisms with opposable thumbs.
(Sapiens.org)

Homo habilis is thought by many to be the first anatomically modern humans, evolving in Africa 2.8 million years ago. These hominids walked upright with hair covering their entire body, and only reached 3.7 feet in height. *Homo habilis* developed the first stone tools used to crush bones and skin animals. This is an example of hominids evolving to use the surrounding landscape to their benefit.
Top - (Encyclopedia Britannica)
Bottom - (Word History Archives)

Homo erectus evolved 2 million years ago and was the first hominid to live outside the African landmass, with fossils found in Europe and Asia. These hominids had much larger brains than their predecessors, and even grew up to 6 feet in height. We can see a massive step up from *homo habilis* in the span of just a few hundred thousand years.
(Encyclopedia Britannica)

Homo heidelbergensis is a link between our cousins, the Neanderthals and Denisovans. The image shows that these hominids started to use fire, and were even able to cook their meat, leading to better brain growth.
(Science Photo Library)

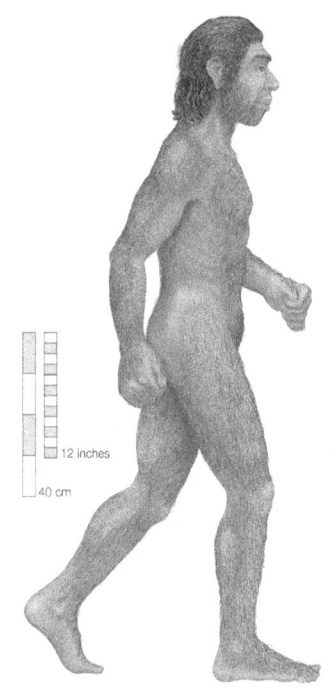

Homo neanderthalensis are an extinct group of humans that lived 400,000 years ago. These hominids had robust and stockier builds, with wider pelvises and shorter legs, reaching an average height of just 5.5 feet for males and 5.1 feet for females. The Neanderthals went extinct around 40,000 years ago.
(Encyclopedia Britannica)

Homo sapiens first emerged 350,000 years ago in Africa, and soon migrated to colonise the rest of the landmass on Earth. Have a look back and see how we started off from the smallest arthropods in the ocean, to eventually evolve into the image on the right.
(Science Photo Library)

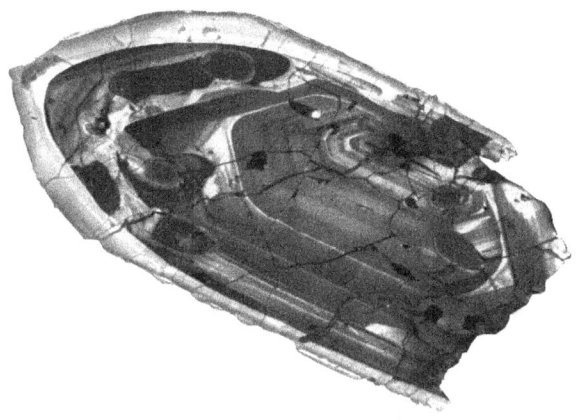

The image above shows a very zoomed in view of a zircon crystal that was found in the Jack Hills of Western Australia. This crystal is said to be 4.4 billion years old, making it the oldest terrestrial rock ever found on Earth.
(National Geographic)

An image of Big Bertha, the moon rock taken from the lunar surface during the Apollo 14 mission. It is said to be terrestrial in origin, with an estimated age of 4 billion years.
(Lunar Sample Laboratory Facility)

An image of the Murchison meteorite, one of the oldest materials ever found on Earth. It has an age of 7 billion years, making it older than the Sun and solar system itself. It has silicon carbide particles in it, as well as a handful of amino acids and alkanes.
(Natural Museum of Natural History)

An example of rock strata, horizontal layers of sedimentary rock that have been formed over millions of years. Each layer represents a distinct period of geological time. The rocks, minerals, and fossils in each layer are studied to provide a better understanding of Earth's history.
(Science Photo Library

Glossary

Aetas Terrae
The age of the Earth.

Alpha Particle
A particle consisting of 2 protons and 2 neutrons with a positive change of +2. Identical to a helium-4 nucleus.

Angiosperms
Flowering plants with ovaries surrounded by a protective fruit.

Angular Rotation
How fast an object rotates or revolves with respect to its centre of rotation. Also known as angular velocity.

Antihydrogen
The antimatter counterpart of hydrogen, made up of an antiproton and a positron.

Antimatter
Matter composed of antiparticles. An example is antihydrogen.

Antiparticles
A particle with the same mass as an ordinary matter particle, but with opposite electric charge.

Antiproton
The negatively charged antiparticle of the proton.

Antiquark
The antiparticle of a quark. Each of the six quark flavours has its own antiparticle, or antiquark.

Aphelion
The point in the orbit of a celestial body that is furthest away from the Sun.

Asteroid
A small, irregular shaped, rocky object that orbits the Sun.

Astrobiology
The study and search of life in the universe. It covers the origins, evolution, and future of life in the universe, based on a number of factors.

Astrochemistry
The study of the abundance and reactions of elements and molecules in the universe, and their interaction with radiation.

Astronomy
The study of everything in the universe beyond Earth's atmosphere. Includes celestial bodies, nebulae, and galaxies.

Astrophysics
A science that employs the methods and principles of physics, chemistry, and mathematics, in the study of astronomical objects and phenomena.

Australopithecus
A very early hominid that existed in Africa, and one of the first ancestors of humans capable of walking upright.

Glossary

Avemetatarsalia
A clade of reptiles more closely related to birds. The two most successful groups were the dinosaurs and the pterosaurs.

Baryon
A type of composite subatomic particle that contains an odd number of quarks, such as the proton and neutron. All baryons are fermions as they have half-integer spins.

Beta Particle
A high energy, high speed, electron or positron emitted by the radioactive decay of an atomic nucleus during beta decay.

Big Bang Nucleosynthesis
The production of atomic nuclei other than hydrogen-1, such as helium-3, helium-4, deuterium, and lithium-7, during the early phases of the universe.

Big Bang Theory
The idea that the universe began with and expanded from a miniscule dense collection of energy called a singularity.

Biochemistry
The study of molecular structures, chemical reactions, and the compositions of substances within living organisms.

Bioturbation
The disturbance of sedimentary deposits by living organisms.

Bipedalism
A form of terrestrial movement where a four-limbed organism moves using its rear limbs, or walks on two feet.

Black Hole
A region of space that has a gravitational field so intense that no matter or radiation can escape.

Blueshift
The shift in wavelength of light towards the blue end of the electromagnetic spectrum, as an object moves towards us.

Boson
A subatomic particle whose spin number has a whole integer value. Most bosons act as force carriers.

Brown Dwarf
Substellar objects that have more mass than the biggest gas giant planets, but less mass than main-sequence stars. Not counted as a star since it can't sustain nuclear fusion.

Centrifugal Force
An outward force that acts on an object that is moving in a circular path. Usually opposes the gravitational force which acts inwards.

Chandrasekhar Limit
The maximum mass of a stable white dwarf star, 1.4 solar masses. If a white dwarf exceeds this limit, it undergoes a supernova reaction and forms either a neutron star or black hole.

Cohesion
The sticking together of particles of the same substance, usually applied to water particles.

Comet
A celestial object made of ice and dust that orbits the Sun, and has a 'tail' of gas and dust when near the Sun.

Coriolis Effect
An inertial force that acts on objects in motion and rotates them with respect to an inertial frame. Responsible for weather patterns.

Cosmic Inflation
The faster-than-light, exponential expansion of the universe during the first fractions of a second after the Big Bang.

Cosmic Microwave Background
Cooled remnant of the first light that could ever travel freely throughout the universe, 380,000 years after the Big Bang.

Cosmic Rays
High-energy particles from outer space that travel throughout the universe at close to light speed.

Cosmic Year
Time taken for the solar system to make one complete orbit around the centre of the Milky Way. It takes about 225 million years.

Cosmos
Another word for 'universe', first used by Pythagoras in 530 BC.

Coulomb Barrier
The energy barrier that two nuclei need to overcome so that they can get close enough to undergo a nuclear reaction.

Craton
An old and stable part of the Earth's continental crust, now the foundation of modern day continents.

Deuterium
An isotope of hydrogen with 1 proton and 1 neutron.

Diffraction Grating
An optical element that divides light of many wavelengths into individual wavelengths, travelling in different directions.

Doppler Shift
The change in frequency of a wave in relation to an observer who is moving relative to the source of the wave.

Dwarf Planet
A spherical celestial object that orbits the Sun but does not achieve orbital dominance like the eight classical planets.

Eccentricity
How much a curved section deviates from being circular.

Eddy Currents
Closed loops of electric current that are induced by a changing magnetic field.

Electromagnetic Force
An interaction between electrically charged particles via an electromagnetic field.

Glossary

Electron
A negatively charged subatomic particle found in all atoms, belonging to the lepton family of elementary particles.

Electron Degeneracy Pressure
The compression of electrons into a tiny volume that halts the collapse of a white dwarf if it is below the Chandrasekhar limit.

Electron Antineutrino
The antiparticle of the electron neutrino, emitted in β^- decay.

Electron Neutrino
An elementary particle with zero electric charge, belonging to the lepton group.

Electroweak Theory
A theory that describes the unification of the weak force and the electromagnetic force.

Elementary Particle
A subatomic particle that is not composed of other particles and cannot be broken down into smaller constituents.

Endothermic
A chemical reaction where the reactants absorb more heat from the surroundings than is released.

Eukaryotes
All organisms whose cells have a membrane-bound nucleus.

Euxinia
Occurs when water contains no oxygen and an increased level of hydrogen sulphide.

Exoplanet
Any planet beyond our solar system.

Exothermic
A chemical reaction where the reactants release more heat to the surroundings than is absorbed.

Extraterrestrial
Any object or life that does not originate from the Earth.

Fermion
A subatomic matter particle that has a half integer spin, includes all quarks, leptons, and baryons.

Frost Line
The minimum distance from a central star where volatile compounds can condense into solid grains.

Galaxy
A system of stars, stellar remnants, nebulae, gas, dust, and dark matter, bound together by gravity.

Gamma Ray
A highly penetrating form of radiation produced as a byproduct of alpha and beta decay.

Gauge Boson
An elementary bosonic particle that acts as a force carrier for elementary fermion particles.

General Relativity
Describes how the four-dimensional fabric of spacetime is warped and curved by mass, and how this curvature causes gravity.

Geocentric
The idea that the Earth is the centre of the universe.

Geochronology
The science of determining the age of rocks, fossils, and sediments using radioactivity.

Geology
The scientific study of the Earth's structure, composition, and many processes, by investigating rocks and sediments.

Gluon
A massless elementary particle that mediates the strong force between quarks by acting as an exchange or 'messenger' particle.

Grand Unified Theory (GUT)
A model in particle physics that merges the strong, weak, and electromagnetic force into a single force at high energies.

Gravitational Constant
An empirical physical constant used in the calculation of the gravitational attraction between two objects.

Gravitational Potential Energy
The potential energy an object has due to its position in a gravitational field.

Gravitational Waves
Ripples in the fabric of spacetime caused by the intense gravity generated by binary stars and merging black holes.

Graviton
A hypothetical elementary particle that mediates or 'transfers' gravitational force. This particle would be a gauge boson.

Gravity
A force that causes mutual interaction between all objects of mass, attracting them towards their centre of mass.

Gymnosperms
Vascular plants that reproduce by means of an exposed seed.

Habitable Zone
The distance from a star where liquid water can exist on a planet's surface.

Hadrons
Any composite subatomic particle made from two or more quarks, held together by the strong force.

Haematite Tubes
Tubelike structures made from an iron ore, haematite, produced by microbes in hydrothermal vents more than 4 billion years ago.

Half-Life
The time taken for a sample of radioactive atoms to reduce, or decay, by half of its original amount.

Glossary

Heliocentric
An astronomical model in which the planets revolve around the Sun, which is at the centre of the universe.

Higgs Boson
An elementary particle with no spin or electric charge, that is present in the Higgs field which gives mass to all other particles.

Intergalactic Medium
Matter and radiation that exists in between galaxies.

Interplanetary Medium
Matter and energy that fills the space between objects in a star system

Interstellar Medium
Matter and radiation that exists in the space between the stars in a galaxy.

Invertebrate
Animals that do not have a backbone, or any bones at all.

Ionising Power
Particles or radiation that has enough energy to ionise an atom or molecule by removing its electrons.

Isotope
Atoms of the same element with the same number of protons but different number of neutrons in their nucleus.

Kaon
A particle with a whole spin integer, formed by the binding of a quark and an antiquark.

Lepton
An elementary particle with a half spin integer that is not affected by the strong nuclear force, just the weak nuclear force.

Lunar Eclipse
An astronomical event that occurs when the Moon moves into the Earth's shadow, causing it to be darkened.

Lycophytes
A group of spore breeding vascular plants, the oldest living terrestrial plants on Earth.

Magnetic Braking
The loss of a star's angular momentum due to its magnetic field.

Meson
A type of composite subatomic particle formed by an equal number of quarks and antiquarks.

Metallicity
The abundance of elements present in a celestial object that are heavier than hydrogen and helium.

Metaphysics
A branch of philosophy that examines the basic structure of reality.

Meteorite
A solid piece of space debris that falls to the surface of a planet or moon.

Microgravity
A condition where the effects of gravity seems to be very small, such as in free-fall or weightlessness.

Glossary

Molecular Cloud

An interstellar cloud that contains absorption nebulae and that can form molecules.

Muon

An elementary particle belonging to the lepton family, with a half integer spin number.

Muon Neutrino

An elementary particle belonging to the lepton family, with zero electrical charge. The muon neutrino produces muons.

Nebulae

Interstellar clouds of gas and dust that are formed from the remnants of dead stars, and are used to form new stars.

Neutrino

The most abundant mass particle in the universe, that only interacts via the weak force and gravity.

Neutron

A composite subatomic particle with no electric change, formed from three quarks (up, down, down).

Neutron Degeneracy Pressure

Occurs in neutron stars when neutrons can no longer be packed any closer, and prevents further collapse of the star.

Neutron Star

The collapsed core of a massive supergiant star.

Nuclear Fission
The process of breaking or splitting large atomic nuclei into smaller nuclei to release large amounts of energy.

Nuclear Fusion
The process in which two light atomic nuclei merge to form a single heavier nucleus, while releasing large amounts of energy.

Nucleosynthesis
The creation of new atomic nuclei from pre-existing nuclei.

Orbital Period
The time taken for an astronomical object to complete one orbit around another object.

Orbital Resonance
Occurs when two astronomical bodies have orbital periods that are expressible in whole number ratios, allowing them to exert a stronger gravitational force on each other.

Paleogeography
Study of the change of Earth's surface, including its landmasses, through time, driven by plate tectonics.

Palaeontology
The study of ancient life by analysing fossil remains.

Panspermia
The hypothesis that life was brought to Earth from space by objects, radiation, or advanced extraterrestrial beings.

Pathogen
Any agent or organism that can produce a disease.

Pauli's Exclusion Principle
States that no two electrons in the same atom can have identical values for all four of their quantum numbers.

Penetrating Power
The ability of each type of radiation to pass through matter.

Perihelion
The point in the orbit of a celestial body that is closest to the Sun.

Philosophy
The study of general and fundamental questions concerning topics such as existence, reason, knowledge, value, and mind.

Phosphorescence
A process in which energy absorbed by a substance is released relatively slowly in the form of light.

Photon
An elementary particle that is a minute packet of electromagnetic radiation, and is the force carrier for the electromagnetic force.

Pion
A composite particle formed from a quark and an antiquark, also the lightest meson.

Planck Length
The smallest possible measurable length where all ideas of gravity and spacetime breakdown, and quantum effects dominate.

Exploring the Essence of Everything

Planetary Migration

Occurs when a planet in orbit around a star interacts with other matter, causing it to move closer or further away from the star.

Planetesimal

A tiny body that combines with many others under gravity, to form a planet.

Plasma

Superheated matter where electrons are ripped away from atoms, causing the whole structure to be ionised.

Plate Tectonics

The scientific theory that explains how major landforms are created as a result of Earth's subterranean movement.

Positron

The positively charged antiparticle of the electron.

Prokaryote

A single-celled organism that does not have a nucleus or other membrane bound organelles.

Proton

A positively charged subatomic particle that occurs in all atomic nuclei, made up of three quarks (up, up, down).

Protoplanetary Disk

A rotating circumstellar disk of dense gas surrounding a newly formed star.

Glossary

Protostar
A young star that is still gathering mass from its molecular cloud or nebula.

Pseudoscience
Statements, beliefs, or practices that claim to be scientific and factual but are incompatible with the scientific method.

Quantum
The minimum amount of any physical entity involved in an interaction.

Quantum Chromodynamics
The theory that describes the action of the strong nuclear force.

Quantum Gravity
A field of theoretical physics that seeks to describe gravity according to the principles of quantum mechanics.

Quantum Field Theory
A theoretical framework that combines quantum mechanics and relativity to explain the behaviour of subatomic particles.

Quantum Fluctuation
The temporary random change in the amount of energy in a point in space, responsible for initial density differences during inflation.

Quantum Mechanics
A field of physics that describes how matter and light behaves at, and below, the scale of atoms.

Quantum Number
Four sets of numbers that are used to describe the position and energy of an electron in an atom.

Quark
An elementary particle with six different types or 'flavours', and a fundamental constituent of matter.

Radioactivity
The release of energy by radiation, from the decay of unstable atomic nuclei.

Radioisotope
A radioactive isotope of an element that is unstable.

Radioisotope Thermoelectric Generator
A type of nuclear battery that converts the heat released from radioactivity into electricity.

Redshift
The shift in wavelength of light towards the red end of the electromagnetic spectrum, as an object moves away from us.

Scalar Boson
A boson that only has magnitude, and whose quantum spin is zero.

Solar Wind
A stream of charged particles that continuously flows outwards from the Sun, in the form of plasma.

Spacetime
A mathematical model that combines three spatial dimensions and one dimension of time into a single four-dimensional continuum.

Spectral Line
A dark or bright line in an otherwise uniform and continuous spectrum that results from the emission or absorption of light.

Spectroscopy
The study of the absorption and emission of light from the electromagnetic spectrum by atoms and molecules.

Spin Quantum Number
A number that describes the angular momentum of a particle. Half-integer values for all fermions and whole integer values for all bosons.

Standard Model
The theory that describes three of the four fundamental forces and classifies all known elementary particles.

Star
A self-luminous gaseous spheroid which produces energy by means of nuclear fusion reactions.

Steady State Model
The theory that the density of matter in the expanding universe remains unchanged due to a continuous creation of matter.

Stellar Nucleosynthesis
The creation of atomic nuclei by nuclear fusion reactions within stars.

Stratification
The layering of sedimentary and igneous rocks in the ground.

String Theory
A theoretical framework in which point-like particles are replaced by one-dimensional objects called strings.

Stromatolite
A layered sedimentary formation built by photosynthetic cyanobacteria.

Strong Force
A fundamental interaction of nature that binds quarks together to form hadrons, and binds protons and neutrons together to form atomic nuclei.

Supernova
The colossal and luminous explosion of the core of a supermassive star as it reaches the end of its nuclear fusion.

Supernovae Nucleosynthesis
The creation of new and heavy atomic nuclei during a supernova.

Surface Tension
The tendency of a liquid surface at rest to shrink into the minimum surface area possible, a sphere.

Synchrotron
A cyclic particle accelerator that accelerates particles using electromagnetic fields, to study the high energy particle collisions.

Tau
An unstable elementary particle of the lepton group with negative electric charge and a half-integer spin.

Glossary

Tau Neutrino
An elementary particle of the lepton group with no electric charge, involved in very high energy nuclear reactions.

Technosignature
A measurable property that provides scientific evidence of past or present technology.

Terraform
The process of modifying a planet, moon, or other celestial body to a more habitable atmosphere, temperature, and ecology so that it can support human life.

Theory of Everything
A final, ultimate theory, that would unite all the forces of nature and that fully explains and links together all aspects of the universe.

Vertebrate
An animal that has a backbone and a skeleton, includes all fish, mammals, reptiles, amphibians, and birds.

Volcanism
The eruption of molten rock from inside an astronomical body to its surface.

W Boson
A vector boson that carries the weak force and that can change the identity of a particle. Found in beta decay.

Weak Force
One of the four fundamental forces of nature, responsible for the radioactive decay of atoms.

Wormhole
A hypothetical structure connecting disparate points in spacetime.

Xenobiology
A subfield of synthetic biology that attempts to design forms of life with different biochemistry and genetic codes than on Earth.

Z Boson
A vector boson that carries the weak force.

Further Reading

Chapter 1

1. *A Brief History of Time* by Stephen Hawking
2. *A Universe From Nothing* by Lawrence M. Krauss
3. *Astrophysics for People in a Hurry* by Neil deGrasse Tyson
4. *At the Edge of Time* by Dan Hooper
5. *Cosmos* by Carl Sagan
6. *Edwin Hubble: Mariner of the Nebulae* by Gale E. Christianson
7. *Flashes of Creation* by Paul Halpern
8. *Genesis of the Big Bang* by Ralph Alpher and Robert Herman
9. *How the Universe Got Its Spots* by Janna Levin
10. *In Search of the Big Bang* by John Gribbin
11. *Kepler* by Max Casper
12. *The Alchemy of the Heavens* by Ken Croswell
13. *The Beginning and the End of Everything* by Paul Parsons
14. *The Cosmic Microwave Background: How It Changed Our Understanding of the Universe* by Rhodri Evans
15. *The Creation of the Universe* by George Gamow
16. *The First Three Minutes* by Steven Weinberg
17. *The Realm of the Nebulae* by Edwin Powell Hubble
18. *Three Degrees above Zero* by Jeremy Bernstein

Chapter 2

1. *A Star Called the Sun* by George Gamow
2. *Antimatter* by Frank Close
3. *Deep Down Things: The Breathtaking Beauty of Particle Physics* by Bruce A. Schumm
4. *Elementary Particle Physics* by Andrew J. Larkoski
5. *Modern Particle Physics* by Mark Thomson
6. *Niels Bohr's Times: In Physics, Philosophy, and Polity* by Abraham Pais
7. *Obsessive Genius: The Inner World of Marie Curie* by Barbara Goldsmith
8. *Quantum: Einstein, Bohr and the Great Debate About the Nature of Reality* by Manjit Kumar
9. *The Atom: A Visual Tour* by Jack Challoner
10. *The Grand Design* by Stephen Hawking and Leonard Mlodinow
11. *The Internal Constitution of the Stars* by Arthur Eddington
12. *The Large Hadron Collider: The Extraordinary Story of the Higgs Boson and Other Stuff That Will Blow Your Mind* by Don Lincoln
13. *The Man Who Changed Everything: The Life of James Clerk Maxwell* by Basil Mahon
14. *The Particle at the End of the Universe: How the Hunt for the Higgs Boson Leads Us to the Edge of a New World* by Sean Carroll
15. *The Theory of Almost Everything* by Robert Oerter
16. *Six Easy Pieces* by Richard Feynmann
17. *Strange Glow: The Story of Radiation* by Timothy J. Jorgensen
18. *Understanding the Atom* by Isaac Asimov

Further Reading

Chapter 3

1. *A Brief History of Earth* by Andrew H. Knoll
2. *A Short History of Nearly Everything* by Bill Bryson
3. *Annals of the Former World* by John McPhee
4. *Archeology: Theories, Methods, and Practice* by Colin Renfrew
5. *Dinotopia: A Land Apart from Time* by James Gurney
6. *Dragon Teeth* by Michael Crichton
7. *Essentials of Geology* by Stephen Marshak
8. *Footprints of Thunder* by James F. David
9. *Human Origins* by New Scientist
10. *Jurassic Park* by Michael Crichton
11. *Krakatoa* by Simon Winchester
12. *Otherlands: A World in the Making* by Thomas Halliday
13. *Raptor Red* by Robert T. Bakker
14. *The Ends of the World* by Peter Brannen
15. *The Fifth Beginning: What Six Million Years of Human History Can Tell Us about Our Future* by Robert L. Kelly
16. *The Lost World* by Arthur Conan Doyle
17. *The Rise and Fall of the Dinosaurs* by Stephen L. Brusatte
18. *The Sixth Extinction* by Elizabeth Kolbert
19. *The Story of Earth: The First 4.5 Billion Years, From Stardust to Living Planet* by Robert Hazen
20. *Trilobite! Eyewitnesses to Evolution* by Richard Fortey
21. *Understanding Earth* by John P. Grotzinger
22. *When Life Nearly Died* by Michael J. Benton

Index

0-9

60 Years Near the Telescope 15

A

absorption line 146, 147, 164
absorption nebula 43
accretion disk 64, 73
Aeneid 173
AI 20
Alexander the Great 8
alien 2-4, 6, 19, 33, 107, 205, 222, 233
Alighieri, Dante 173
Allosaurus 200, 250
Alpha Centauri 37
alpha particle 104, 128, 132, 136-41, 163
alpha process 128
Alpher, Ralph 31, 123, 159
americium-241 139
amino acid 152, 259
amniotes 192, 203
Andromeda 37
Angels & Demons 114
angiosperm 201, 204
angular momentum 73, 81
Ankylosaurus 200, 250
anomalocaris 239
antibaryon 116
antihydrogen 115, 119
antimatter 101, 114-20, 148, 158
antiparticles 115, 118
antiproton 115, 118, 154
antiquark 115, 118
Apatosaurus 250
aphelion 79, 228
Apollo 180, 218, 258
arachnids 190, 241
Archaeopteryx 200, 248
Archean 170, 179-85, 188
archeology 140, 168, 213
argon-40 215
Aristotle 8, 9
arthropleura 193, 245
arthropod 188-92, 203, 201, 239, 257
Asimov, Isaac 5, 15, 16, 107
asteroid 9, 12, 41, 51, 82, 174, 180, 204, 219, 226
asthenosphere 176
astrobiology 7, 9, 10
astrobotany 9
astrochemistry 17, 18, 82, 232
astronaut 3, 12, 15, 46, 65, 218
astronomy 9, 13, 28, 31, 77
astrophysics 17, 26, 117
Atopodentatus 198
ATP 141-42
Attenborough, David 6
australopithecus 207

Avalonia 190
Avemetatarsalia 198

B
Baltica 190
Banner, Bruce 133
baryon 98, 105, 114, 116
basalt 181
Bell Telephone Laboratories 27
beryllium-8 128
beryllium-10 214
beta minus decay 104-06
beta plus decay 104-06
Becquerel, Henri 134-35, 162, 211
Bethe, Hans 123-25, 160
Big Bang 7, 20, 24-7, 29-32, 54, 64, 82, 96, 112-18, 122-26, 131, 148, 164, 204
Big Bang nucleosynthesis 121-26, 159
Big Bertha 218, 258
biochemistry 15, 95
bioturbation 189
black dwarf 58, 66
black hole 11, 14, 18, 37, 45, 63-7, 112, 118, 226
Blade Runner 5
blueshift 25, 147
Bohr, Neils 137, 146, 164,
boson 96-99
Brachiosaurus 200, 220, 249

Brown, Dan 114
brown dwarf 58
Burbidge, Eleanor 130, 160
Burbidge, Geoffrey 130, 160

C
C. Clarke, Arthur 5
Cambrian 187-89, 239
Cambrian explosion 187, 236
carbohydrates 95, 99
carbon-12 128, 131, 215
carbon-14 140, 213-15
carbon dating 99, 140, 213-15
carbon dioxide 176-78, 183-85, 192, 196, 245
Carboniferous 187, 191, 243
Carnivora 206
Cenozoic 171, 203-05
centrifugal force 48, 49, 174
cephalopod 239
CERN 101, 114, 118, 119, 150
Chandra X-ray Observatory 45
Chandrasekhar limit 61
Chariots of the Gods 2
Chicxulub crater 204
chlorophyll 142
chloroplast 142
Clerk Maxwell, James 109, 156
Cloud, Preston 173, 236
CNO cycle 126, 127
cobalt-60 213
Coelophysis 198, 246

Columbia 186
comet 43, 50-52, 79,
Compton, Arthur 153
Compton effect 153
consciousness 16, 19, 82
constellation 3, 43
Contact 10
continental crust 178, 181, 229
continental drifting 186
Cooksonia 241
Copernicus 7, 77, 221, 222
corona 40, 127, 230
Cosmic Calendar 6, 36, 54, 59, 168, 182, 204
cosmic dust 42
Cosmic Microwave Background 26, 29-32, 82, 109
cosmic rays 40, 118, 133, 158
cosmic year 36
Cosmos 7, 10
Coulomb barrier 125
Crab Nebula 90
cratons 181, 185-86
Creodonts 206, 254
Cretaceous 198-202, 205, 249
Cruise, Tom 178
Curie, Frèdèric 135
Curie, Irene 135
Curie, Marie 9, 135, 136, 162, 211
Curie, Pierre 135, 162
curium 132
cyanobacteria 181, 187-89, 194, 238

D
da Vinci, Leonardo 9, 169
dark nebula 42, 43, 88, 89
Darwin, Charles 170, 233
David 195
David Anderson, Carl 118, 158
Deimos 48, 51
deGrasse Tyson, Neil 1
Democritus 4
Denisovans 209, 256
deuterium 58, 107, 122-24
Devonian 187, 190-92, 242
diamictite 184
Dicke, Robert 29-31
diffraction grating 144
diffuse nebula 42-44, 89
dinosaur 82, 193, 195-205, 219, 246
Diplodocus 200
Dirac, Paul 117, 118, 158
DNA 139, 206, 230
doppler shift 24
Drake Equation 20
Dune 5
dwarf planet 76
dystopian 5

E
eccentricity 78-80
Eddington, Arthur 124, 160
Einstein, Albert 9, 26, 79, 110, 115, 221
einsteinium 132

electromagnetic force 25, 97, 101, 104, 108-110
electrometer 135
electron 31, 62, 96, 105
electron antineutrino 105, 106, 115
electron degeneracy pressure 61
electron neutrino 97, 106
electron pump 107
electron volt 143
electroweak theory 108
elementary particle 95-99, 102, 118, 150-55
ellipse 78
elliptical orbit 78, 79, 228
emission line 147
emission nebula 42, 43, 89
endothermic 129
energy level 142-47
Epicurus 4
Epsilon Indi 37
Eris 76
eukaryote 184
Euramerica 190, 191
Europa 10
euxinia 196
event horizon 64, 87
evolution 17, 42, 72, 113, 148, 170, 181, 184, 188, 192-4, 203, 210, 223
exoplanet 4, 10, 17, 19, 41, 58, 78, 147, 222, 233

exothermic 129
extinction 2, 43, 82, 183, 196-99, 202-05, 208, 210, 220
extraterrestrial 2, 3, 9, 10, 19, 29

F
Faraday's Law of Induction 109
felsic basaltic rock 181
Fermi, Enrico 155
Fermi's Paradox 20
fermion 96-99
fermium 132
first light 31, 32
Ford, Harrison 168
fossil 168, 188, 211, 215-220
Foster, Jodie 10
Fowler, William 130, 161
FOXP2 gene 209
Fraunhofer lines 145, 164
frost line 74, 75

G
galactic centre 37, 86, 225
galaxy 26, 33, 37-41, 54, 223
galaxy filament 38, 39
galaxy formation 24, 26, 33
Galileo 9
gamma ray 32, 66, 126, 138-140
Gamow, George 31, 123, 124, 159

Gamow factor 124, 159
Gargantua 65
gas giant 74, 75
gauge boson 97
Gauss 229
Geiger, Hans 136
Gell-Mann, Murray 151
general relativity 26, 157
genetic mutation 139, 140
geocentric 77, 221
geochronology 170
geology 140, 168-70, 212
German Geological Society 185
Gliese 65
gluon 96-99
glyptodonts 208
God Equation 15
Goldilocks zone 226
Goliath 195
Gondwana 189, 191, 199
Gorgonopsia 145
GPS 11
Graham Bell, Alexander 28
Grand Unified Theory 113
granite 191, 218
gravitational constant 111, 112
gravitational force 25, 35, 47-51, 63, 80, 97, 102, 113
gravitational perturbations 180, 226
gravitational potential energy 57
graviton 97, 110
Great Carbonation Event 183
Great Oxidation Event 182

gymnosperm 193, 201, 240

H

habitability 41, 49, 50
habitable zone 41
Hadean 170, 177, 178, 204, 217
Hades 173
hadron 98, 121
haematite tubes 220
half-life 212-216
Halley, Edmond 79
Halley's Comet 79
Hanks, Tom 114
Heinlein, Robert 9
Heisenberg, Werner 146, 165
heliocentric 77, 221
Helios 147
helium 42, 47, 54, 61, 76, 123-29
Hera 36
Hercules-Corona Borealis 39
Herman, Robert 31
Hess, Victor 118
hexapods 190
Higgs boson 94, 97
Higgs field 97
Holmdel Horn Antenna 28, 85
Holmes, Arthur 169, 170, 234
Holocene 210
hominid 209
homo erectus 208, 209, 256
homo ergaster 208
homo habilis 207, 255
homo heidelbergensis 209, 256

homo neanderthalensis 210, 257
homo sapiens 6, 204, 209, 233
Hoyle, Fred 24, 124, 130, 161
Horsehead Nebula 88
Hot Jupiter 80, 81
Hubble, Edwin 25, 26, 84
Hulk 134
Huronian glaciation 184
Hutton, James 169, 170, 234
Hyde Wollaston, William 144
hydrogen 32-35, 43-45, 60, 72, 156
hydrothermal vents 231
hylonomous lyelli 244
Hyperion 174

I

Ibn al-Haytham 8
Ibn Rushd 8
Ibn Sina 8, 169
Ice Age 208
Ichthyosaur 197-202, 247
igneous rock 181
Illuminati 114
inertia 80
Inferno 173
inflation 24, 26, 27, 32-34, 86
INTEGRAL 132
intergalactic medium 45
intergalactic space 40, 45
interplanetary space 39-41, 45

Interstellar 5, 14, 64
interstellar medium 21, 41-45, 55
interstellar space 40, 42, 130
invertebrate 189, 193
ionise 43, 44, 61
ionising power 138, 139
iron 60, 75, 126-132, 176, 182, 196
iron-nickel alloy 229
infrared ray 126, 143
inner core 175, 229
isotope 58, 122, 130,
ISS 12, 46

J

JPL 11
Jupiter 19, 41, 50, 58, 74, 80-82, 174, 180, 221
Jurassic 199, 249
Jurassic Park 5, 168, 200

K

K. Dick, Phillip 5
Kaku, Michio 6, 15
kaon 99
Kepler, Johannes 77-80, 91, 221
kinetic energy 63, 174, 177
King Philip 8
King, Stephen 6
Kuiper belt 174
Kuo, Shen 169

L
Lane, Solomon 178
Langdon, Robert 114
Laniakea Supercluster 38
Large Hadron Collider 101, 150
Late Cambrian 139
Late Carboniferous 191
Late Heavy Bombardment 179-181
Late Paleozoic 246
Late Paleozoic Ice Age 192
Laurentia 190
Lehmann, Inge 235
Lemaître, Georges 24, 25, 29, 84
lepton 96-99, 106, 117
light speed 37, 62, 126, 138, 150
light year 26, 37-39
LIGO 132
lissamphibian 193, 197
lithium 123-125, 132
Local Group 37, 38
lunar eclipse 77
lutetium-177 140
lycophytes 197
Lyell, Charles 169, 170, 235

M
Macedon 8
Maillard reaction 13
magnesium 129, 131, 176
magnetic braking 50
magnetic field 41, 50, 87, 118, 229-231
main sequence stars 67, 68, 72
Mammalia 203, 204
Manhattan Project 160
mantle 176, 181, 184, 229-231
marine biology 14
Mars 9, 15, 19, 48, 77, 82, 174, 180, 221, 230
Marsden, Ernest 136
Martian 2, 3, 5, 6, 12, 15, 232
Marty 213, 214
McConaughey, Mattthew 10, 65
meganeura 193, 245
Mercury 50, 51, 60, 75, 82, 226, 227
meson 98, 99
Mesozoic 171, 196, 197, 200
Messier 78
metallicity 44, 225
metaphysics 20
meteor 199, 202
meteoroid 50, 52
methane 184
microbial mats 187, 189
microfossils 220
microgravity 20
microorganism 13, 140, 187, 194
Milky Way 36, 37, 48, 49, 63, 73, 86, 87, 223
Milton, John 173
Minority Report 5
Mission Impossible: Fallout 178

moschops capensis 244
molecular cloud 43, 49, 54-59, 73
molecular gastronomy 13
mollusc 188-190, 239
moon 10, 19, 48, 50-52, 76, 218, 225
Moon 51, 174, 175, 218, 228, 229
Mosasaur 201, 205, 251
muon 97
muon neutrino 97
Murchison meteorite 218, 219, 259
myriapods 190

N
NASA 12, 45, 119
Nath Bose, Satyendra 152
Neanderthals 209, 256, 257
nebula 11, 18, 24, 43, 50, 54, 66
Neptune 51, 74, 81, 174, 180, 226
Neogene 206
neon-20 129
neuroscience 14
neutrino 126, 127
neutron 96-104, 120-22, 130-33,
neutron degeneracy pressure 61, 63
neutron star 63-67, 74, 118, 132

Newton, Isaac 9, 14, 55, 104, 144, 221
Newton's Law of Gravitation 110
nickel 130, 175, 176, 229
Nicomachus 8
nitinol 12
nitrogen 125-127, 140, 175
Noah 179
Nolan, Christopher 14, 65
Nothosaurus 198, 199
nuclear fission 138, 148
nuclear fusion 11, 43, 50-59, 122, 138, 148
nuclear physics 26
nucleosynthesis 121-132, 159, 161

O
On the Origin of Species 170
orbit 76-78, 81, 91, 102, 224
orbital period 78-80
orbital resonance 81
Ordovician 187, 189, 240
Orion 43
Orion Nebula 43, 87
Ørsted 109
Ortelius, Abraham 185, 237
osteoporosis 12
outer core 176, 229

oxygen 60, 62, 95, 124-128, 175, 177, 181, 183, 186, 189, 196, 232
ozone layer 127, 231, 232

P

palaeontology 195, 199, 202, 219
Pale Blue Dot 18
Paleogene 205, 254
paleogeography 191, 205
Paleozoic 171, 187, 190, 192-194
Paradise Lost 173
Panem 5
Pangaea 185, 191, 193, 199, 205
Pangaea proxima 205
panspermia 219
Panthalassa 193
parabolic orbit 78
pathogens 2
pathology 14
Pauli exclusion principle 109
penetrating power 138, 139
Penzias, Arno 28-32, 85
perihelion 79, 226
Permian 187, 193, 201-203, 245, 246
Permian-Triassic extinction 196
Phanerozoic 170, 187, 196
Philemon Coleman, Arthur 184, 236
Phobos 48, 51

phosphorescence 134, 162
photon 25, 31-34, 109, 116, 120, 138, 142-147
photosynthesis 141, 143, 148, 181, 183, 189, 192
pion 99, 155
Pioneer 10
Pisces-Cetus Supercluster 238
placoderm 190
Planck length 26
Planck, Max 153
planet V 180
planetary migration 76, 80, 81
planetary motion 77, 80, 91
planetary nebula 42, 44, 62, 65, 66, 89
planetary system 18
planetesimal 74
Plato 8
plectronoceras sp. 239
Plesiosaur 198-201, 247
pleuromeia 197, 246
plum pudding model 148
Pluto 75, 76, 79
plutonium 175
polonium 135, 136
positron 106, 115-119, 126, 127, 138, 158, 215
potassium-40 214-216, 240
primates 206, 207, 254
Princeton University 29
Principia Mathematica 157
Principles of Geology 235
prokaryote 183

Prolia 12
Positron Emission Tomography 119
proteobacteria 183
Proterozoic 170, 182-89, 197, 203
proton 62, 99, 102, 110, 122, 126-132, 163
proton-proton chain 125-127
protoplanetary disk 73-75, 81
protostar 57-59, 73, 74
Proxima Centauri 37, 224
pterosaurs 193, 198-200, 206

Q

quantum chromodynamics 102
quantum field theory 34
quantum fluctuation 33, 35, 37, 82, 86
quantum gravity 26, 64
quantum mechanics 113, 153
quantum theory 26, 165
quark 96-105, 116, 151
quark-gluon plasma 27, 96, 101, 114
Quaternary 208
quartz 218
Quetzalcoatlus 202, 252

R

r-process 130-132

radiation 28-32, 43, 58, 61-63, 66, 72, 123, 133-139, 229, 230
radioactive decay 104, 138, 148, 214, 231
radioisotope 107, 140, 175, 230
radium 135, 136, 140
red giant 60, 62
red supergiant 59-61, 67
redshift 24, 25, 30, 147
reflection nebula 42, 43, 88, 89
Ring Nebula 89
rocky planet 74, 81, 82
Rodinia 186
Röntgen, Wilhelm 134, 135, 162
Royal Society of Edinburgh 234
Rubbia, Carlo 154
Rutherford, Ernest 136, 137, 163
rutherfordium 132

S

s-process 131
sabre-tooth tiger 208, 213, 225
Sagan, Carl 6, 10, 23, 224
Sagittarius A* 36, 87
satellite 7, 9, 11, 19, 28, 51
Saturn 19, 50, 74, 81
sauropsid 192, 244
scalar boson 97
scattering 32, 43

Schrödinger, Erwin 117, 146, 158, 165
sci-fi 3, 5, 14, 16, 20, 63
scientific method 9, 14
Scott, Ridley 15
sedimentary rock 184, 219, 259
seismology 235
Selene 174
semi-major axis 78
SETI 4, 10, 17
Shelley, Mary 5
silicates 176, 184
silicon 129, 259
Silurian 187, 190, 241
Simulation Theory 20
singularity 25, 64
Slipher, Vesto 24, 29
snowball Earth 189
solar mass 58, 60, 61, 63, 70, 127
solar radius 70
solar system 13, 19, 37, 51, 53, 75, 78-82, 112, 168, 180, 218, 226
solar wind 40, 41, 52, 74, 230
spacetime 25, 33, 35, 63, 64, 82, 157, 232
spectroscopy 142, 143
Spider-Man 134
spin number 96
Spinosaurus 202, 253
Stanford Linear Accelerator 101, 151

star system 36-39, 45, 60, 225-228
starstuff 67
Staurikosaurus 198
steady state model 29, 31
Stegosaurus 199, 200, 249
stellar nebula 54
stellar nucleosynthesis 125, 126, 161
stellar nursery 43
stellar wind 50
strata 168, 206, 219, 259
string theory 101
stromatolites 187-189, 194, 210, 220, 238
sulphur 129, 196
Super Proton Synchrotron 154
supermassive black hole 87
supernova 42, 44, 62, 66, 74, 130
supernova remnant 42, 44, 66, 67, 90
supernovae nucleosynthesis 121, 130, 131
surface tension 46, 47
synapsid 192, 193, 244

T
Tarter, Jill 93
tau 97
tau neutrino 97
technosignature 19
tectonic plates 176, 181, 185, 194

telescope 9, 19, 32
temnospondyl 197
tetrapod 191, 192, 242
tetrapodomorphs 191
Titan 10
The Divine Comedy 173
The Dragons of Eden 7
The Fly 5
The Gods Themselves 107
The Martian 5, 15
The Matrix 176
The Penultimate Truth 176
The Science of Interstellar 65
The Star Beast 9
The War of the Worlds 3
Theia 174, 175, 218
Theory of Everything 113
Theory of the Earth 169
theropods 200
Thesaurus Graphicus 185
Thomson, J.J. 136, 152
thorium 175, 230
Thorne, Kip 14, 65
tidal lock 81, 227
Tikhov, Gavriil 9, 14
tiktaalik 242
TRAPPIST-1 224
Triassic 196-99, 246, 247
Triceratops 202, 253
Trifid Nebula 89
Trigonotarbida 241
trilobite 190, 240

triple alpha process 128
tritium 122
Tyrannosaurus rex 199, 202, 251

U

UFO 3
ultraviolet rays 43, 127, 142, 177, 182, 231
uranium 135-38, 163, 175
uranium-232 230
uranium-235 230
uranium rays 134
Uranus 51, 74, 81, 180, 227

V

Van Allen radiation belt 230
van der Meer, Simon 154
vascular plants 190, 191, 197
Velociraptor 199, 202, 252
Venus 10, 51, 60, 78, 82, 221, 227
Verne, Jules 5, 176
vertebrate 190-193, 198-200, 206, 244, 247, 252
Virgil 173
VIRGO 132
Virgo Supercluster 38, 39
volcanism 180, 200, 231
von Daniken, Erich 2
von Fraunhofer, Joseph 144, 164

Voyager 10

W
W boson 106, 107
weak force 25, 27, 101, 104-110, 137, 155
Wegener, Alfred 185, 237
Weir, Andy 5, 15
Wells, H.G. 3, 5
white hole 64
white dwarf 65-67
Whitman, Walt 167
Witton, Mark 202
woolly mammoth 208
Woodrow Wilson, Robert 28-32, 85

X
X-rays 134, 135, 162
xenobiology 9
Xenophanes 169

Y
yttrium-90 140
Yukawa, Hideki 155

Z
Z boson 97, 106, 154,
zircon 218, 258
Zweig, George 151

www.ingramcontent.com/pod-product-compliance
Lightning Source LLC
Chambersburg PA
CBHW050050230526
45470CB00004B/1469